Social Tourism at the Crossroads

Social tourism is at a pivotal point. Over the past decade, there has been increased interest and research into social tourism issues, and as a result there is now much greater evidence on the important role that social tourism can play in providing significant benefits for the people supported through social tourism schemes. However, despite these advances and awareness of the benefits of tourism participation in most countries, there is still much confusion and ambiguity about the definition, role, and purpose of social tourism.

This comprehensive volume reflects recent shifts in social tourism research by focusing on target groups and the benefits or constraints of these groups in holiday participation. The authors explore the diversity of issues, theories, and social contexts that are relevant to social tourism research, offering a range of quantitative and qualitative methods and experimental designs as well as various policy and practice contexts to address policy issues. They also highlight opportunities for greater intensity of research on the importance of policy in advancing social tourism and to stress the fundamental role that social tourism can play in achieving strategic policy goals towards enhancing wellbeing, citizenship, and quality of life in the future.

This book was originally published as a special issue of the *Journal of Policy Research in Tourism, Leisure and Events*.

Anya Diekmann is Professor of Tourism at the Université Libre de Bruxelles, Belgium. Her research integrates aspects of social tourism in Belgium and Europe with a focus on social tourism systems and the benefits of tourism on well-being and social inclusion. Anya is co-ordinator of the Alliance for Research of ISTO (International Organisation of Social Tourism).

Scott McCabe is Professor of Marketing and Tourism, and Associate Dean for Research, at the Nottingham University Business School, UK. His research focuses on the qualities of tourist experience, consumer behaviour, and tourist decision-making. He is currently editor-in-chief (alongside Sara Dolnicar) of *Annals of Tourism Research*.

Carlos Cardoso Ferreira is an Assistant Professor in the Department of Geography and Tourism at the University of Coimbra, Portugal. His main research areas are tourism planning and development, ageing and old age mobilities, and social and inclusive tourism. He has authored several publications on tourism and ageing, tourism development and planning, and senior tourism.

Social Tourism at the Crossroads

Edited by
**Anya Diekmann, Scott McCabe and
Carlos Cardoso Ferreira**

Routledge
Taylor & Francis Group

LONDON AND NEW YORK

First published 2020 by Routledge

2 Park Square, Milton Park, Abingdon, Oxon, OX14 4RN
605 Third Avenue, New York, NY 10017

Routledge is an imprint of the Taylor & Francis Group, an informa business

First issued in paperback 2020

British Library Cataloguing in Publication Data
A catalogue record for this book is available from the British Library

ISBN 13: 978-0-367-25817-7 (hbk)
ISBN 13: 978-0-367-78474-4 (pbk)

Typeset in Minion Pro
by RefineCatch Limited, Bungay, Suffolk

Publisher's Note
The publisher accepts responsibility for any inconsistencies that may have
arisen during the conversion of this book from journal articles to book chapters,
namely the inclusion of journal terminology.

Disclaimer
Every effort has been made to contact copyright holders for their permission to
reprint material in this book. The publishers would be grateful to hear from any
copyright holder who is not here acknowledged and will undertake to rectify
any errors or omissions in future editions of this book.

Contents

Citation Information

The chapters in this book were originally published in the *Journal of Policy Research in Tourism, Leisure and Events*, volume 10, issue 3 (November 2018). When citing this material, please use the original page numbering for each article, as follows:

Chapter 1

Editorial – Social tourism: research advances, but stasis in policy. Bridging the divide
Anya Diekmann, Scott McCabe and Carlos Cardoso Ferreira
Journal of Policy Research in Tourism, Leisure and Events, volume 10, issue 3 (November 2018), pp. 181–188

Chapter 2

Social tourism as a modest, yet sustainable, development strategy: policy recommendations for Greece
Konstantinos I. Kakoudakis and Scott McCabe
Journal of Policy Research in Tourism, Leisure and Events, volume 10, issue 3 (November 2018), pp. 189–203

Chapter 3

'Not on my vacation': service encounters between able-bodied and disabled consumers – the case of high-contact service
Anat Tchetchik, Victoria Eichhorn and Avital Biran
Journal of Policy Research in Tourism, Leisure and Events, volume 10, issue 3 (November 2018), pp. 204–220

Chapter 4

The holiday-related predictors of wellbeing in seniors
Marlène Mélon, Stefan Agrigoroaei, Anya Diekmann and Olivier Luminet
Journal of Policy Research in Tourism, Leisure and Events, volume 10, issue 3 (November 2018), pp. 221–240

Chapter 5

Accessible tourism and its benefits for coping with stress
Andreia Filipa Antunes Moura, Elisabeth Kastenholz and Anabela Maria Sousa Pereira
Journal of Policy Research in Tourism, Leisure and Events, volume 10, issue 3 (November 2018), pp. 241–264

Chapter 6

Chapter 7

For any permission-related enquiries please visit:
http://www.tandfonline.com/page/help/permissions

Notes on Contributors

Stefan Agrigoroaei is a Professor at the Psychological Sciences Research Institute at the Université catholique de Louvain, Belgium.

Avital Biran is a Senior Lecturer and Principal Academic in Tourism Management in the School of Tourism at Bournemouth University, UK. Her research interests revolve around issues of consumer behaviour and experience in tourism and leisure, including psychological and cross-cultural aspects.

Carlos Cardoso Ferreira is an Assistant Professor in the Department of Geography and Tourism at the University of Coimbra, Portugal. His main research areas are tourism planning and development, ageing and old age mobilities, and social and inclusive tourism. He has authored several publications on tourism and ageing, tourism development and planning, and senior tourism.

Eugenia Devile is a Lecturer at Coimbra Superior School of Education (ESEC) at the Polytechnic Institute of Coimbra, Portugal. Currently, she is the Director of Tourism degree and a member of ESEC Scientific Council.

Anya Diekmann is Professor of Tourism at the Université Libre de Bruxelles, Belgium. Her research integrates aspects of social tourism in Belgium and Europe with a focus on social tourism systems and the benefits of tourism on well-being and social inclusion. Anya is co-ordinator of the Alliance for Research of ISTO (International Organisation of Social Tourism).

Victoria Eichhorn is Professor of Tourism at the Business School at Hochschule Fresenius – University of Applied Sciences, Germany. Her research is multi-disciplinary focusing on exclusionary practices in tourism.

Pilar Espeso-Molinero is a Post-doctorate Faculty Member in the Faculty of Humanities and Philosophy at the University of Alicante, Spain. She teaches Cultural Heritage and Tourism Anthropology while co-ordinating the Social Anthropology Department.

Claire Haven-Tang is the Research Degree co-ordinator for the Cardiff School of Management and a Reader in Tourism and Management at the Welsh Centre for Tourism Research at Cardiff Metropolitan University, UK.

Konstantinos I. Kakoudakis is a Lecturer in Hospitality and Tourism Management at UCLan Cyprus, Cyprus. His research interests revolve around social tourism, sustainable development, and working conditions in the Hospitality industry. He was previously Researcher/ Consultant in tourism development projects co-funded by the EU, while he also has several years of work experience in the tourism industry. He holds a PhD from the University of Nottingham, UK.

Elisabeth Kastenholz is Associate Professor in Marketing, Consumer Behaviour, and Tourism in the Department of Economics, Management, Industrial Engineering and Tourism (DEGEIT), and a member of the Research Unit on Governance, Competitiveness and Public Policies (GOVCOPP), at the University of Aveiro, Portugal.

Olivier Luminet is a Professor at the Psychological Sciences Research Institute at the Université catholique de Louvain, Belgium, and is also affiliated with the Belgian Fund for Scientific Research (FRS-FNRS), Belgium.

Scott McCabe is Professor of Marketing and Tourism, and Associate Dean for Research, at the Nottingham University Business School, UK. His research focuses on the qualities of tourist experience, consumer behaviour, and tourist decision-making. He is currently editor-in-chief (alongside Sara Dolnicar) of *Annals of Tourism Research*.

Marlène Mélon is a Research Analyst at the Psychological Sciences Research Institute at the Université catholique de Louvain, Belgium. She has specialised in (neuro)psychology of the elderly and the cognitive processes involved in Alzheimer's disease.

Andreia Filipa Antunes Moura is an Adjunct Professor at the Coimbra Superior School of Education (ESEC) at the Polytechnic Institute of Coimbra, Portugal, and an Integrated Researcher at the Research Unit on Governance, Competitiveness and Public Policies (GOVCOPP), at the University of Aveiro, Portugal.

Anabela Maria Sousa Pereira is an Associate Professor with Aggregation in the Department of Education and Psychology (DEP), and a member of the Research Centre on Didactics and Technology in the Education of Trainers (CIDTFF), at the University of Aveiro, Portugal.

Diane Sedgley is a Senior Lecturer in Tourism in the Department of Tourism, Hospitality and Events, and a member of the Welsh Centre for Tourism Research at Cardiff Metropolitan University, UK.

Anat Tchetchik is a Senior Lecturer in the Department of Geography and Environment at Bar Ilan University, Israel. She researches economic geography.

Social tourism: research advances, but stasis in policy. Bridging the divide

Introduction

In the last 10 years, there has been a significant increase in interest in, and research on social tourism issues, particularly in the European context (McCabe, Minnaert, & Diekmann, 2011; Minnaert, Maitland, & Miller, 2009). Consequentially, there is now much greater evidence on the important role that social tourism can play in providing significant benefits for the people supported through social tourism schemes. These benefits have also been shown to extend to wider society and to destinations, businesses, governments and communities, which welcome social tourists and their spending. This research is building and spreading beyond Europe, to include a wide range of countries, supporting a diversity of social objectives, to the extent that we now have a much greater knowledge on the extent of social tourism across the world and the important roles it plays in societies.

For example, recent research has conceptualised the links between social tourism and sustainable development (McCabe, 2018). Other studies have addressed the needs of specific social tourist segments in a range of countries, such as elderly people in Slovenia (Gabruc, Medaric, & Sedmak, 2018), or have addressed the transferability of social tourism systems to developing economies, such as Kazakhstan (Assipova & Balnur, 2017). Yet others have noted the important types of benefits felt by specific groups, such as the educational benefits of group tours of older Mexican people (Estrada-González, 2017) or how governments in South America developed social tourism policies (de Almeida, 2011; Schenkel, 2013). We have learned of the role that social tourism has played in the apartheid and post-apartheid era social life in South Africa (Adinolfi & Ivanovic, 2015), and an analysis of the potential for inbound social tourism in Egypt (Hamad, 2017). Whilst the research emerging may be characterised as nascent, exploratory or focusing on a particular aspect of social tourism, it is without doubt that there has been widespread, inter-disciplinary interest in the concept worldwide.

Social tourism policies emerged in many countries in Europe and the Americas since the introduction of paid holidays for workers (Jolin & Proulx, 2005). These policies focused in general on programmes for specific target groups or on the public funding for social tourism provision. This policy context has remained fairly stable over a long period and has not witnessed much development over time. Yet in recent years, some governments have renewed their involvement in the area, which signals perhaps a growing demand for social tourism amongst the population, or a growing awareness by policy makers of the emerging evidence of social tourism's potential to deliver benefits to local populations (either as consumers or producers of social tourism services). For example, a new nationwide initiative, called 'ScotSpirit', was launched by Visit Scotland, the national tourism board for Scotland in 2016 (see www.visitscotland.org). Whilst this is still at the small scale, and was supported by the London-based, U.K.-wide charity, the Family Holiday Association, that has been campaigning to put Social Tourism on the political agenda in the U.K. for many years, it represents a significant milestone. Another example of policy development is the recently revised policy on 'tourism for all' by the Flemish tourism organisation Toerisme Vlaanderen in 2015 (https://www.vakantieparticipatie.be). This policy provides access to holiday participation for all, based on the recognition that tourism is a fundamental right for all citizens, and that everyone should be able to enjoy holidaymaking, and due to its contribution as a core asset for mental and physical wellbeing.

However, other governments with long-established social tourism programmes have cut existing aids, due to budget constraints, as for example Greece and Spain where governments severely reduced public funding, which supports holiday participation. In Portugal, the Inatel project, which in the past provided widespread support for senior citizens to access a holiday, has seen its funding completely ceased since 2011. However, in 2018, Inatel relaunched this flagship, publicly funded programme.

Yet, despite these advances and awareness of the benefits of tourism participation in most countries, there is still much confusion and ambiguity about the definition, role, and purpose of social tourism. Some research conflates social tourism with social enterprise, for example (Franzidis, 2018), and generally, the research lacks a sustained, and systematic focus, so there is a need for more research and greater awareness building amongst the academic community as well as public bodies. Thus, whilst there have been important advances, the socio-political and economic landscape of the last 10 years has been dominated by the effects and aftermaths of the Global Financial Crisis. Social tourism programmes have been hit by budget cuts, which have never recovered to pre-2007 levels, and yet there is evidence of new and hybrid schemes emerging. Therefore, at the level of research there is a great deal to celebrate, however, in terms of the effects of the research on policy and practice, there is limited success. Hence, the need for a closer examination of the policy issues in social tourism.

Emerging themes in social tourism research

Social tourism has been defined as comprising;

> ... all activities, relationships and phenomena in the field of tourism resulting from the inclusion of otherwise disadvantaged and excluded groups in participation in tourism. The inclusion of these groups in tourism is made possible through financial or other interventions of a well-defined and social nature. (Minnaert, Diekmann, & McCabe, 2011, p. 29)

Whilst the definition of social tourism presented here is not universally accepted, it does have many of the elements that are shared by other researchers working in this field. Generally, social tourism concerns an inclusionary agenda, such that it has strong links to the 'Tourism For All' concept. The latter has originally been developed to specifically address the needs of people living with disabilities. Whilst we might argue that the two concepts are entirely distinct, they share many similar characteristics in terms of exclusionary issues and practices (see for example: Eichhorn, Miller, & Tribe, 2013).

Recent research has also noted the huge potential challenges and scale of disability and access issues for people with disabilities, as well as the profoundly emotional and meaningful experiences of these tourists and their carers (Lehto, Luo, Miao, & Ghiselli, 2017). Yet whilst social tourism is primarily driven by an agenda of inclusion, it also concerns a wider set of issues other than accessibility. In that inclusionary perspective, the Flemish government introduced the term Tourism for All to replace the (often misunderstood) term social tourism, since social tourism deals with a wide spectrum of constraints and barriers, economic as well as social and physical. In addition, social tourism can be used as a stimulus to demand, which can be used as part of the counter-seasonal strategies of destinations (Cisneros-Martínez, McCabe, & Fernández-Morales, 2018), indeed one of the key arguments in the EU preparatory action Calypso (European Commission, 2011).

Social tourism has thus evolved from a concept designed to facilitate holiday participation amongst low-income groups following the introduction of paid holidays for workers in the 1930s and 1940s. It is now a much broader, holistic concept focusing on disadvantaged groups in society with the goal of fostering inclusion, social mixing and citizenship, notably through holiday participation. The core value behind the social tourism concept is the idea that that 'having a break' from daily life (and problems) contributes to the social, mental, and physical wellbeing of all individuals and subsequently contributes to 'good' health (Diekmann & McCabe, 2016). From that perspective, social tourism is driven by a goal to provide consumption opportunities with an added moral value to all members of society (Minnaert et al., 2009). The increasing research and recognition of the significant contribution of tourism participation to quality of life and health issues as a component of wellbeing (McCabe & Johnson, 2013; McCabe, Joldersma, & Li, 2010), has very likely contributed to its position and integration within the social systems in many European countries (Diekmann & McCabe, 2011).

Yet, it should be underlined that the inclusionary aims of social tourism are somewhat antagonistic with the focus on target groups of social tourism (families, youngsters, seniors, disabled) confirmed by the European preparatory action Calypso in 2009. Policy makers and governments often prefer to develop policies aiming at these specific groups disregarding the fundamental social tourism value of inclusion.

Correspondingly research shifted from definitions and practices in Europe towards to a more socio-political approach, looking into the benefits of holiday participation and its relationship with wellbeing and health (for a summary, see Smith & Diekmann, 2017). Studies highlighted the difficulty of determining one broad ranged beneficiaries' category as impacts on consumption, benefits and practices are subject to strong variations according to age. In that perspective, research focuses more and more on the relatively under-researched target groups and the circumstances or conditions that constrain different groups from being able to fully engage in tourism and what the different groups gain from holiday participation. In addition, in recent years research emerged tackling tourism's role in individual's lives and serving to broaden the level of interest in social tourism as a field of investigation by other disciplines, such as psychological and medical sciences looking into psychological and physiological effects of tourism on consumers (Diekmann & McCabe, 2016).

This brings us to the purpose of the current special issue, the focus of which is on the policy implications and issues surrounding social tourism. Yet, the research seems to have little impact on policy and knowledge transfers and exchange (KTE) between research and policy makers is very limited. This is even the case when research is funded by governmental agencies, but then do not necessarily acknowledge the research or utilise them for building new or integrating into existing policies. The reasons are multiple, such as change in the make-up of authorities after elections, for instance, budget restrictions, personal interests and priorities of politicians and policymakers, etc.

However more significantly, the focus of such an enterprise is bound to be on the evidence. Having worked in the field of social tourism for many years, the editors of this issue understand the problem of evidence. In order to make policy, there must be a basis in 'evidence'. This is a problem that needs to be addressed. How can we ensure that the research has maximum impact on policy and decision making amongst key stakeholders, how can we insure knowledge transfer (KTE)?

A wide literature on KTE related to health – and thus close to the social tourism debate – highlights the various difficulties that exist to ensure knowledge transfer, such as researchers and decision makers being driven by demands that may not be conducive to successful KTE, as adapting and justifying activities that are not in line with traditional academic expectations (Mitton, Adair, McKenzie, Patten, & Perry, 2007, p. 730). Another cause advanced by

Lavis, Robertson, Woodside, McLeod, and Abelson (2003) consists in the lack of transferring 'actionable messages from a body of research knowledge, not simply results of a single study' (p. 223). There are certainly myriads of possibilities to be developed by both sides to improve communication, such as for researchers, in addition to research results, to deliver more recommendations to policy makers.

However, research-based 'evidence' is emerging at a faster rate, across contexts and is becoming theoretically richer and methodologically more rigorous. Yet, while in recent years most research has used qualitative methods, useful to understand the benefits for tourism on the individuals' wellbeing (mental and physical health) and tackling in/exclusion, advances in quantitative research might be needed to deliver figures that evidence the economic benefits of tourism participation. So far, only very little research – along with some grey literature (e.g. assessment studies of specific social tourism programmes such as those managed by Imserso and Inatel) – has looked into the cutting of health expenditures after holiday participation (Eusebio, Carneiro, Kastenholz, & Alvelos, 2013). More 'figures orientated' research is needed for building a solid argumentative to inform policy makers.

Conclusions

In summary, a shifting of the focus of the evidence, towards an exploration of the benefits to society, as opposed to the individual, which has been the main concern of research in recent years. Secondly, there is a need to shift the focus from the short-term to the long term. We know that there will be an explosion in the proportions of older people in society, living with all manner of disabilities. This demographic time-bomb needs an industry that can respond to the access, information and services requirements of an elderly and 'infirm' population in order to stay relevant and in business. The important role that policymakers have in ensuring also the focus on the long-term objectives of a sector cannot be underplayed. Businesses tend to be focused on the expediencies of the short-term need for profit, and yet policymaking is also increasingly being driven by the same expediencies.

Social tourism is at a pivotal point. Therefore, there is an urgent need for further research at the policy level that seeks to interrogate the issues facing the social tourism sector in the future and which addresses the interdisciplinary and inter-sectoral challenges. This special issue is only the second that is dedicated to social tourism research. Most of the papers have been presented at the ISTO Congress in Zagreb at the meeting of the Alliance on Training and Research in Social and Fair Tourism on the 19th October 2016. The Alliance, created in 2010 by the international organisation of social tourism (ISTO) (http://www.oits-isto.org/oits/public/section.jsf?id = 181), started several years ago to reunite social tourism researchers for a bi-annual international conference in order to exchange the latest research, to build knowledge in social tourism and in consequence to inform policy makers. While there is a real demand for KTE within ISTO members composed by policy makers and various social tourism stakeholders, up to now, academics and practitioners have held their sessions separately. Acknowledging, however, the urgent need to debate social tourism requirements and changes in demand, the 2018 international congress in Lyon will favour exchange and debates between research and practitioners across various sessions.

The contents of this special issue

The papers in this special issue reflect the above-mentioned shift in social tourism research as five of them focus on target groups and the benefits or constraints of these groups in holiday participation. All papers offer a diverse range of methods, quantitative, qualitative, and experimental designs as well as a wide range of policy and practice contexts to address policy issues. Each paper provides recommendations for policymakers, thus shedding a new light on the future paths social tourism has ahead.

The first paper, by Kakoudakis and McCabe, provides a novel contribution through a critical analysis of social tourism policies/programmes in Greece. This provides a detailed evaluation of how social tourism policies have been shaped by the geo-political and economic forces in recent years, thus adding further to our understanding of the uneven development of social tourism in different countries, even in similar geo-political contexts (e.g. EU). From research and evidence on the benefits of these programmes in other countries the authors refer to the potential of social tourism to act as stabilising force in the Greek tourism system, vis-à-vis the actual socioeconomic environment that characterises Greece, in the aftermath of the 2008 crisis. The paper also argues that while the potential of social tourism has not been fully untapped it can play a pivotal role towards the broader goal of sustainable development as it addresses some of the major socioeconomic problems of Greece, such as unemployment, acute poverty, the shrinking of SMEs, and social exclusion.

In the second paper, Tchetchik, Eichhorn, and Biran develop an interesting and novel approach to the issues of the perceptions of disability within a tourism context for consumers. Whilst we might imply that huge strides have been made within society in terms of the 'normalisation' of disability through positive role models and portrayals of people with disabilities in mainstream media, there is still a long way go in terms of the inclusionary agenda in a tourism context. There are numerous examples of studies which have found unconstructive or negative attitudes amongst tourism and hospitality service personnel towards customers with disabilities. And yet, there are also many examples of businesses that have explicitly sought to integrate disability-perspectives and images into their service offerings and marketing. However, there is a real gap in the literature since most studies focus on the views and experiences of people who have disabilities, and whilst this is of course extremely important, there are few studies that have attempted to obtain the views of able-bodied guests who share the service environment with disabled guests. In order to explore the attitudes of able-bodied towards the presence of people with disabilities in shared holiday space, the authors develop an interesting stated preference choice experiment. The findings show that there is a long way to go before tourists are fully accepting of the presence of disabled guests. This is a matter of societal concern and hence, the need for policy interventions.

The study presented by Melon et al. details the Wallonian government-funded research, which focused amongst other things, on the impact of holiday participation on mental and physical wellbeing. Until recently, only a few studies have investigated this association in seniors and the impact of holiday-related predictors of wellbeing by comparing travellers and non-travellers, particularly on how holidays impact on social and cognitive activities and the degree to which perceived health benefits were associated with higher wellbeing. A sample of 4130 seniors participated in a survey of people's last holiday, daily activities, health, and wellbeing. Results showed that senior tourists were younger, more educated, wealthier, and healthier than senior non-tourists. In addition, the levels of wellbeing were higher in senior tourists compared to senior non-tourists, after adjusting for control variables. The results also showed that frequent holidays, a greater frequency of social and cognitive activities, as well as the degree of perceived health benefits were associated with higher levels of wellbeing. This quantitative analysis provides original and novel data, adding significantly to a still poorly developed research area. In that perspective, the outcomes not only highlight the importance of holidays to maintain psychological and physical health, but the research also provides the basis for a discussion on the development of new policies.

In a similar research with individuals with disabilities, though more focused on stress management, Moura, Kastenholz and Pereira show how accessible tourism helps to cope with stress. The authors take, as in the research of Melon et al. a psychological perspective by adapting the Leisure Coping Scale applying it to the Accessible Tourism context and analysing the individuals' biopsychosocial dimensions of stress-coping. The results of the research again deliver evidence for the positive influence of holidays on life balance, health and wellbeing for travellers with disabilities, particularly in the context of rehabilitation. It also provides a systematisation of the biopsychosocial dimensions that are positively influenced by accessible leisure tourism, thus emphasising the potential of tourism as an effective policy to promote social inclusion, health and wellbeing of individuals with disabilities. The research informs policymakers and suggests integrating new policies of alternative therapeutic interventions and develops new tourism products targeted to a population with special needs.

A further study on the issues surrounding people with disabilities is presented in the paper by Devile and Kastenholz. This paper focuses on the concerns and needs of people with visual impairments. It identifies how the literature on constraints, facilitators, and strategies to negotiate them have evolved in the literature. This is important since it better reflects the experiences of people with disabilities as their conditions and life circumstances are not static but are adjusted and contextually specific. The authors call for a more holistic approach to understanding the perceptions and feelings of people with visual impairments, through a qualitative interview method. The findings identified a whole range of negative attitudes and behaviours experiences by people with visual disability. Visual impairments are treated as a homogenous category and people perceived erroneous views amongst able-bodied people about the issues facing them. However, the important implications from these findings are in the negotiation strategies used, and in the ways that visually impaired tourists coped with constraints and identified a range of facilitators. Whilst these are positive steps in the right direction, the paper concludes with important implications for policy and regulation, and particularly the need to enforce regulations for people with disabilities, to truly progress towards a more inclusive tourism industry.

Haven-Tang, Sedgley, and Espeso-Molinero address one of the oldest and flagship social tourism initiatives: the Spanish Government's IMSERSO programme. They allude to the limited attention given to this topic in regards to older people, considering it an under-researched group in relation to the personal and social benefits of social tourism initiatives. The authors take a demand-side perspective supported by in-depth interviews to analyse older people's engagement and how social tourism is used as part of a holistic approach to older people's welfare. Social tourism is hereby described as an innovative public policy approach that should accompany other conventional European models of support for older people enhancing their subjective wellbeing and social inclusion.

Taken together, the papers in this special issue highlight the diversity of issues, theories and social contexts that are relevant to the field of social tourism research. The research demonstrates the policy relevance of social tourism, including the interconnectedness of different policy spheres. Whereas tourism is often positioned in relation to industrial policy and strategy, what is striking about social tourism is that it can and should crosscut policy areas including; health, social welfare and industry policy. Indeed, there are a range of other policy areas including education and equality and diversity and others for which social tourism can also contribute. This special issue hopes to highlight opportunities for greater intensity of research on the importance of policy in advancing social tourism and to stress the fundamental role that social tourism can play in achieving strategic policy goals towards enhancing wellbeing, citizenship, and quality of life in the future.

References

Adinolfi, M. C., & Ivanovic, M. I. L. E. N. A. (2015). Recounting social tourism development in South Africa. *African Journal for Physical Health Education, Recreation and Dance, 21*(Supplement 2), 1–12.

Assipova, Z., & Balnur, A. (2017). Investigation of using Belgian case of social tourism in Almaty, *Critical Tourism Studies Proceedings*. Article 134. Retrieved from http://digitalcommons.library.tru.ca/cts-proceedings/vol2017/iss1/134

Cisneros-Martínez, J. D., McCabe, S., & Fernández-Morales, A. (2018). The contribution of social tourism to sustainable tourism: A case study of seasonally adjusted programmes in Spain. *Journal of Sustainable Tourism, 26*(1), 85–107.

de Almeida, M. V. (2011). The development of social tourism in Brazil. *Current Issues in Tourism, 14*(5), 483–489.

Diekmann, A., & McCabe, S. (2011). Systems of social tourism in the European Union: A comparative study. *Current Issues in Tourism, 14*(5), 417–430.

Diekmann, A., & McCabe, S. (2016). Social tourism and health. In M. Smith. & L. Puczkó (Eds.), *The Routledge Handbook of Health Tourism* (pp. 103–112).

Eichhorn, V., Miller, G., & Tribe, J. (2013). Tourism: A site of resistance strategies of individuals with a disability. *Annals of Tourism Research, 43*, 578–600.

Estrada-González, A. E. (2017). Social tourism, senior citizens, and education. *World Leisure Journal, 59*(Suppl. 1), 22–29. doi:10.1080/16078055.2017.1393872

European Commission. (2011). Calypso means tourism for all. DG Enterprise and Industry.

Eusebio, C., Carneiro, M. J., Kastenholz, E., & Alvelos, H. (2013). The socioeconomic impacts of INATEL social tourism programmes for the senior market. In A. Diekmann, & L. Jolin (Eds.), *Regards croisés sur le tourisme social dans le monde* (pp. 197–213). Quebec: Presses de l'Université du Québec.

Franzidis, A. (2018). An examination of a social tourism business in Granada, Nicaragua. *Tourism Review*, doi:10.1108/Tr-04-2017-0076

Gabruc, J., Medaric, Z., & Sedmak, M. (2018). Social tourism for seniors in Slovenia. *Revista Turismo & Desenvolvimento, 2*(27/28), 39–41.

Hamad, M. M. A. (2017). Social tourism in Egypt: Travel agencies' perspective. *International Journal of Heritage, Tourism, and Hospitality, 10*(2/2), 337–343.

Hunziker, W. (1951). *Social tourism: Its nature and problems*. Geneva: International Tourists Alliance Scientific Commission.

Jolin, L., & Proulx, L. (2005). L'ambition du tourisme social: un tourisme pour tous, durable et solidaire!. Revue Interventions économiques. *Papers in Political Economy*, (32), 1–12.

Lavis, J. N., Robertson, D., Woodside, J. M., McLeod, C. B., & Abelson, J. (2003). How can research organizations more effectively transfer research knowledge to decision makers? *The Milbank Quarterly, 81*(2), 221–248.

Lehto, X., Luo, W., Miao, L., & Ghiselli, R. F. (2017). Shared tourism experience of individuals with disabilities and their caregivers. *Journal of Destination Marketing & Management, 8*, 185–193.

McCabe, S. (2018). Social tourism and its contribution to sustainable tourism. *Cuadernos Económicos de Información Comercial Española, 93*, 29–44.

McCabe, S., & Johnson, S. (2013). The happiness factor in Tourism: subjective well-being and social tourism. *Annals of Tourism Research, 41*, 42–65.

McCabe, S., Joldersma, T., & Li, C. (2010). Understanding the benefits of social tourism: Linking participation to subjective well-being and quality of life. *International Journal of Tourism Research, 12*(6), 761–773.

McCabe, S., Minnaert, L., & Diekmann, A. (Eds.). (2011). *Social tourism in Europe: Theory and practice (Vol. 52)*. Bristol: Channel View Publications.

Minnaert, L., Diekmann, A., & McCabe, S. (2011). Defining social tourism and its historical context. In S. McCabe, L. Minnaert & A. Diekmann (Eds.), *Social tourism in Europe: Theory and practice* (pp. 18–30). Channelview.

Minnaert, L., Maitland, R., & Miller, G. (2009). Tourism and social policy: The value of social tourism. *Annals of Tourism Research, 36*(2), 316–334.

Mitton, C., Adair, C. E., McKenzie, E., Patten, S. B., & Perry, B. W. (2007). Knowledge transfer and exchange: Review and synthesis of the literature. *The Milbank Quarterly, 85*(4), 729–768.

Schenkel, E. (2013). El turismo social como política estatal en Sudamérica. *Pasos. Revista de Turismo y Patrimonio Cultural, 11*(1), 173–183.

Smith, M. K., & Diekmann, A. (2017). Tourism and wellbeing. *Annals of Tourism Research, 66*, 1–13.

Anya Diekmann

Scott McCabe
🆔 http://orcid.org/0000-0002-9807-9321

Carlos Cardoso Ferreira

Social tourism as a modest, yet sustainable, development strategy: policy recommendations for Greece

Konstantinos I. Kakoudakis and Scott McCabe ⓘ

ABSTRACT

Recent findings from social tourism research and evidence from its practice have shown that social tourism has multiple benefits, both for individual participants and host-communities within destinations. The latter in particular have been acknowledged by the European Union and have been included in its recent sustainable tourism policy. Yet, there are a limited number of studies that have attempted to explicate the close linkages between social tourism and sustainable development, and to inform public policy. This paper aims to do so with specific reference to Greek social tourism programmes. Drawing upon development theory, specifically sustainable development, and sustainable tourism research in particular, the study builds an argument for the potential of social tourism to act as a stabilising force in the Greek tourism system, contributing to the achievement of sustainability outcomes for host-communities. In doing so, the paper makes tangible policy recommendations, which are also timely, given the current socioeconomic environment that has been shaped in Greece, across Europe, and elsewhere, since the 2008 crisis.

RESUMEN

Los resultados recientes de la investigación en turismo social y la evidencia de esta práctica han mostrado que el turismo social tiene múltiples beneficios, tanto para los participantes individuales como para las comunidades anfitrionas en los destinos. Esto último en particular ha sido reconocido por la Unión Europea y ha sido incluido en su reciente política de turismo sostenible. Aún así, hay un número reducido de estudios que hayan intentado explicar las estrechas conexiones entre el turismo social y el desarrollo sostenible así como informar la política pública. Este trabajo aspira a esto con referencia específica a los programas griegos de turismo social. Haciendo uso de la teoría del desarrollo, específicamente desarrollo sostenible, y la investigación en desarrollo sostenible en particular, este trabajo presenta un alegato sobre el potencial del turismo social para actuar como una fuerza estabilizadora en el sistema turístico griego contribuyendo al logro de los resultados sostenibles para las comunidades anfitrionas. De esta forma, este trabajo hace tangible las recomendaciones políticas, que son al también oportunas, dado el entorno socioeconómico actual que ha sido moldeado en Grecia, en Europa y en todo el mundo desde la crisis de 2008.

RÉSUMÉ

Les résultats des recherches récentes sur le tourisme social et les données probantes générées par l'aspect pratique ont montré que le tourisme social présente de multiples avantages, tant pour les participants individuellement que pour les communautés d'accueil dans les destinations. Les avantages des destinations ont été particulièrement reconnus par l'Union européenne et ont été intégrés dans sa récente politique du tourisme durable. Néanmoins, quelques études ont tenté d'expliquer l'existence de liens étroits entre le tourisme social et le développement durable, en vue d'influencer la mise en place des politiques publiques. L'analyse de ces liens est le but de la présente étude dont les références sont faites spécifiquement aux programmes du tourisme social grec. S'appuyant sur la théorie du développement, particulièrement le développement durable et la recherche sur le tourisme durable, cette étude confirme que le potentiel du tourisme social est capable de stabiliser le système touristique grec en contribuant à l'acquisition des résultats durables pour les communautés d'accueil. Partant, cet article formule des recommandations concrètes des politiques qui sont également opportunes, eu égard à l'environnement socio-économique actuel qui s'est développé depuis la crise de 2008 en Grèce, en Europe et ailleurs.

摘要

于最近社交旅游的研究及相关证据中，发现个体参与者及当地社区双方都能够受益于社交旅游。对于当地社区的得益更受到欧盟 (EU) 确认，并被纳入可持续旅游政策的一部份。然而，现在只有屈指可数的学术研究在尝试阐明社交旅游与可持续发展旅游两者之间的紧密关系。本研究文章正以此为目标，透过希腊的社交旅游计划来勾画并发展出一套理论，一套特别以可持续发展及持续旅游为研究对象的理论。此学术研究正是建基于这套立论：社交旅游的潜在可能性充当着希腊旅游体系的稳定支柱，并为当地社区带来可持续性收益。为此，本研究文章以现实的政策为参考助证，而亦相当恰时，因为 2008 年环球金融危机造就了现今希腊和欧洲各地的社会经济环境。

Introduction

Increasing evidence from the practice of social tourism programmes in continental Europe (e.g. Spain and Portugal) has shown that social tourism can contribute to the generation of employment and to the economic growth of host-communities that suffer from seasonality (e.g. Cisneros-Martínez, McCabe, & Fernández-Morales, 2017; Eusébio, Carneiro, Kastenholz, & Alvelos, 2013, 2016). These socioeconomic benefits have been acknowledged at the EU level, resulting in social tourism's recognition as an integral part of the EU tourism policy, with the potential to contribute to the achievement of sustainability outcomes for destinations (European Economic and Social Committee [EESC], 2006). On the other hand, in tourism research the concept of social tourism has been mainly studied in terms of the individual psychological benefits for social tourists (e.g. Kakoudakis, McCabe, & Story, 2017; McCabe &

Johnson, 2013). Furthermore, within the sustainable tourism literature in particular, social tourism has been largely neglected. There are only a few studies that have focused on the specific relationship between social tourism and sustainability (e.g. Baumgartner, 2011; Cisneros-Martínez et al., 2017).

As a consequence, existing tourism literature on the socioeconomic linkages between social tourism and sustainability, and the former's potential to contribute to the latter's objectives, remains scarce. This potential in conjunction with the cost-effectiveness of social tourism programmes (see Minnaert, Maitland, & Miller, 2009) make the need for such studies urgent, especially in times of economic recession, following the Global Financial Crisis of 2008, social inequality, and major environmental concerns, phenomena that have taken extreme dimensions, threatening the quality of life of the present, and future, generations. This study attempts to addresses this gap in sustainable tourism literature by focusing on social tourism in Greece, a country that has suffered perhaps more than any other European country from the recent financial crisis and its consequences (Hellenic Statistical Authority [ELSTAT], 2016a).

Tourism, as the main 'industry' of Greece, and thus a major agent of change, has been viewed as the vehicle towards development (Ministry of Economy, Development and Tourism, 2014). Indeed, tourism has been proved to be resilient during this critical period, contributing to a containment of the recession (Foundation for Economic and Industrial Research [IOBE], 2016). On the other hand, the prevailing model of mass tourism comes with major limitations (e.g. seasonality) (Buhalis, 2001), which do not allow tourism as it currently operates to bring about more positive and long-lasting changes in socioeconomic development. This paper argues that social tourism has the potential to address some of these limitations. Drawing upon development theory, specifically sustainable development, and sustainable tourism research in particular, the study builds an argument for the close relationship between social tourism and sustainability outcomes for host-communities at destinations in Greece. In doing so, and given that social tourism has been a neglected concept both in the Greek National Development agenda, and in tourism research in Greece, this paper also aims to stimulate future interest in social tourism amongst policy-makers and tourism scholars.

The paper proceeds with a brief critical review of the literature on sustainable development and tourism, which identifies the neglect of social tourism from these debates, and highlights its inherent linkages with sustainability. Then the current socioeconomic environment of Greece and the role of tourism within this environment are presented, to provide a context for the discussion on the limitations of Greek tourism as it currently operates, and the potential of social tourism to address some of these limitations, and to contribute to the country's socioeconomic stability, and sustainable development. The paper concludes with some tangible policy recommendations that are also transferable to other geographical contexts, with similar socioeconomic characteristics.

Sustainable development and tourism

While the concept of development had been traditionally defined in terms of economic growth, it has evolved over time, mainly due to the failures of political systems to incorporate human well-being (Goulet, 1992) and environmental concerns (Bramwell & Lane,

1993) into their development agendas. This has resulted in the reconceptualisation of development towards a broader definition, which encompasses the multiple dimensions (e.g. economic, social, cultural, environmental, etc.) and complexities of social systems (e.g. poverty, social exclusion, unemployment, guaranteed human rights, and environmental protection), factors that also determine development (e.g. Sharpley, 2000). During this definitional and political shift, the Brundtland Report 'Our Common Future' (World Commission on Environment and Development [WCED], 1987) further contributed to the broadening of the development concept and agenda, by focusing on another parameter of, or requirement for, development, that is, sustainability. The Brundtland Report introduced sustainable development as a new paradigm of development, which 'meets the needs of the present without compromising the ability of future generations to meet their own needs' (WCED, 1987, p. 43). Since then, this new perspective of development has received significant attention in academic and policy circles, but in parallel, it has also been largely misconstrued.

> While sustainable development is intended to encompass three pillars, over the past 20 years it has often been compartmentalised as an environmental issue. Added to this, and potentially more limiting for the sustainable development agenda, is the reigning orientation of development as purely economic growth (UN, 2010, p. 2).

Today, and although misconceptions about the meaning and objectives of sustainable development still exist, it has become clearer that sustainable development reflects a holistic perspective of development, which means 'integrating the economic, social and environmental objectives of society, in order to maximise human well-being in the present without compromising the ability of future generations to meet their needs' (Organisation for Economic Co-operation and Development [OECD], 2001, p. 11). Indicative of this progress, and the increasing emphasis on human well-being, is the development of OECD's 'Better Life' Index (2011), which encompasses new and more holistic sustainable development approaches.

The popularisation of the sustainable development concept within tourism research, has led to the emergence of another concept, that is, sustainable tourism. Sustainable tourism has evolved in parallel with its parental concept, in the sense that it has also been largely misconstrued by tourism scholars. Until relatively recently, debates on sustainable tourism have reified environmental protectionism in the main in light of the climate-change lobby (Liu, 2003; Lu & Nepal, 2009). This has meant an undue emphasis on environmental pillars at the expense of economic, and more importantly, sociocultural dimensions of sustainable development. Today, there has been an increasing recognition amongst tourism researchers that tourism operates within a broader economic, social and physical environment, and that sustainable development is a holistic concept that addresses all three aspects of this environment (McDonald, 2009). As a result, this wider perception of sustainability, and its social pillar in particular, has been addressed in several tourism studies on well-being and the quality of life of residents in host communities (e.g. Chi, Cai, & Li, 2017; Kim, Uysal, & Sirgy, 2013; Woo, Kim, & Uysal, 2015). In line with these advancements in sustainable tourism research, it is imperative for tourism theory and practice to emphasise the potential of tourism to contribute to the well-being and quality of life of all actors involved. This paper argues that the role of social tourism can be pivotal towards this direction.

Social tourism and sustainable development

Social tourism mainly revolves around the inclusion of economically and socially disad-vantaged populations in travel and tourism, through interventions (either policy or finan-cial or other support) of a well-defined social nature (McCabe, Minnaert, & Diekmann, 2011). According to an increasing body of research, this inclusion holds a plethora of sociopsychological benefits for disadvantaged groups (e.g. Kakoudakis et al., 2017; McCabe & Johnson, 2013; Minnaert et al., 2009; Morgan, Pritchard, & Sedgley, 2015). These individual benefits in conjunction with the conceptual foundation of social tourism, may result in the public's perception that social tourism refers exclusively to welfare (e.g. Anglo-Saxon approach). However, evidence from the practice of social tourism in continental Europe (e.g. IMSERSO and INATEL social tourism programmes in Spain and Portugal, respectively) shows that social tourism simultaneously contributes to the generation of employment and to the economic growth of host-communities that suffer from seasonality (Cisneros-Martínez et al., 2017; Eusébio, Carneiro, Kastenholz, & Alvelos, 2016).

Hence, in several countries, such as Spain, Portugal, Belgium, France, and Greece, it has become apparent that the provision of social tourism holidays is much more than welfare. Given the multiple benefits that emanate from its practice, it could be argued that, although the term 'social' hints towards philanthropy, social tourism benefits both individ-ual tourists and the local economies at destinations. In actuality, social tourism fills in the gaps of general or mainstream tourism, namely, its inadequacy in encompassing neglected social groups, and areas in its framework. Moreover, it achieves this in a cost-effective way (Minnaert et al., 2009), while also adopting more environmentally sustainable practices. For instance, social tourism is mainly a domestic form of tourism (thus, resulting in lower carbon dioxide emissions), largely practiced during the off-season, and it often uti-lises more eco-friendly types of accommodation, such as hostels and caravans (see Baum-gartner, 2011; Cisneros-Martínez et al., 2017). In addition, the volume of social tourists is significantly lower than the volume of 'mainstream' tourists, which means that social tour-ism's impact on the carrying capacity of host-communities is rather minimal. Thus, social tourism contributes to several sustainable development indicators, such as public health, social inclusion, socioeconomic development, and sustainable transport (see Eurostat, 2013). As a result of these advantages, social tourism has begun to receive interest at an EU level, and has been viewed as a 'sustainable tourism structure' (EESC, 2006, point 2.4.3). Nevertheless, the emphasis of EU public policy has been on the economic dimen-sion of sustainability (social tourism as a means to manage low-season).

Despite these strong linkages between social tourism and sustainability, within the large volume of sustainable tourism literature, social tourism remains under-researched. This is perhaps due to the fact that the debate on tourism and sustainability has primarily focused on environmental and economic issues, overshadowing socio-cultural ones (Cole, 2006). In one of the rather few studies that have discussed social tourism within the context of sustainability, Ryan (2002, p. 17) stresses that 'in addressing the issues of *sustainability* concepts akin to social tourism need to be applied.' In a similar vein, Higgins-Desbiolles (2006) has advocated the potential of social tourism to deliver various benefits to entire communities. This has been confirmed by recent studies on social tourism programmes in continental Europe. For instance, Eusébio et al. (2013, 2016) found that INATEL social tourism programme for seniors in Portugal had positive effects on local communities, not only in relation to the mitigation of seasonality, but also with regards to diversification and revitalisation of local economies. Similarly, findings from a study by Cisneros-Martínez et al. (2017) on the socioeconomic effects of IMSERSO social tourism programme in Spain show that social tourism is a very useful tool against seasonality, providing greater hotel occupancy in the low-season, which in turn has a positive impact on the wider local market, and can contribute to improved quality of life for residents in host communities.

The above effects are in addition to the positive effects on social tourists, which refer to the inclusion of disadvantaged social groups in tourism participation, and the positive impact of this participation in their quality of life. Hence, these multiple benefits simultaneously address important social and economic issues (e.g. unemployment and social exclusion), which have been persistent in several countries, and especially in those that still suffer from the adverse effects of the recent financial crisis, such as Greece. However, within the Greek tourism literature, and to the best of the authors' knowledge, there is only one conference paper that briefly discusses social tourism in relation to the socioeconomic dimensions of sustainability (see Despotaki, Tsartas, & Doumi, 2015). This is a serious neglect, given the potential of social tourism to respond to some of the current social and economic issues that undermine Greece's sustainable development.

The case of Greece: tourism and the Greek development agenda

Greece makes an interesting case to study these issues for three mains reasons: (1) its current socioeconomic environment is ideal in addressing some major global phenomena (e.g. unemployment, social exclusion, and acute poverty); (2) tourism has been a major agent of change in the country, and one of the few sectors that can contribute to its sustainable development (IOBE, 2016; Ministry of Economy, Development and Tourism, 2014); and (3) although social tourism has formed a part of Greek social policy since the 1980s, and support for its practice has been amongst the main Greek tourism policy targets (Buhalis, 2001; European Commission, 2014), yet it remains a particularly neglected area of study. Hence, Greece does not serve here as a geographical context, but as a socioeconomic context in which tourism plays a central role.

Greece is a country that has suffered severely from the recent financial crisis (ELSTAT, 2016a). Since the beginning of the recession, what has monopolised the rhetoric of the past two Administrations, the policy recommendations of the vast majority of economists, and the headlines of the Greek mass media, is the urgency for Foreign Direct Investment (FDI), which has been viewed as the 'key' to the country's economic recovery and development (Kathimerini, 2016 August). This is in line with the tendency of governments worldwide to approach development from the economic growth perspective (Goulet, 1992). This approach, however, is often not consistent with the aims of sustainable development, including the achievement of long-term benefits for countries and their citizens (OECD, 2001). On the other hand, it could be argued that the focus of Greece on development strategies, such as FDI, has its underlying logic, given that the country has been with its back to the wall for seven consecutive years, and does not have many sectors with the potential to give the necessary boost to its economy.

Arguably, one of the few sectors which comprise the cornerstone of the Greek economy, and, hence, have the greatest potential to contribute to the country's development, is tourism. Greece has been a popular tourism destination, since the expansion of international mass tourism, mainly due to the attractiveness of its ancient history, climate, and natural environment (e.g. Buhalis, 2001). Tourism has a major contribution to Greece's GDP and employment rates, and it is often described as the 'barometer' of its economy. In 2015, for instance, the direct contribution of tourism is estimated at 9.8% of the GDP, and its total contribution (both direct and indirect) at 20–25% (Association of Greek Tourism Enterprises [SETE], 2016). In the same period, tourism had also an important contribution to employment growth (9.0% direct, and 18.9% total contribution) (SETE, 2016). Given the prolonged turbulence in the Greek economy, in the aftermath of a six-year recession, which has severely damaged sectors with a decisive impact on the country's economic and employment growth (e.g. constructions and services), the above figures confirm the resilience of the sector, and its significant contribution to a containment of the recession (IOBE, 2016).

On the other hand, it must be also acknowledged that the recent increase in international arrivals is, to a large extent, attributable to the political turmoil in the Middle East and North Africa, which has strengthened the attractiveness of Greece as a tourist destination. However, tourism demand is never constant, and therefore, destinations cannot take its growth for granted (Liu, 2003). Moreover, and despite this significant boost in its performance, tourism's contribution to the country's socioeconomic stability and sustainable development remains insufficient, especially when considering Greece's financial situation, unemployment rates, poverty, and the immense individual and social implications of these phenomena.

This is not to say that such complex socioeconomic issues is expected to be tackled through tourism alone, but rather to question whether tourism as it has been traditionally operated in Greece has reached its full potential to contribute to their mitigation. The answer is no, mainly due to four major limitations that Greek tourism has: (a) from a tourism perspective, it suffers from seasonality; (b) from an economic perspective, it is managed, to a large extent, by big businesses (e.g. tour-operators, and hotel-chains), which has resulted in the squeezing of small-medium enterprises (SMEs); (c) from a social perspective, it largely excludes those most in need (e.g. low-income families, and unemployed individuals) from receiving the benefits of tourism participation (e.g. psychological benefits); and (d) from an environmental perspective, the prevailing model of mass tourism has largely exploited the country's natural resources (e.g. Buhalis, 2001; ELSTAT, 2016b).

What about the potential of social tourism?

In Greece, social tourism is largely organised and run by public authorities. It was initiated in 1976 by the Greek National Tourism Organisation (EOT), and it was first launched in 1982 by EOT, the Worker's Social Benefits Organisation and the General Secretariat for Youth (Despotaki et al., 2015; European Commission, 2010a). Since then, social tourism programmes have benefited numerous citizens and businesses across the country. Nevertheless, its practice has seen significant fluctuations over time, both in terms of available budget, and volume of beneficiaries. This has been largely due to recurrent changes in public policies, and supply cuts. Indicative of these changes is that in 2012, social tourism programmes were cut down in line with the requirements of 'The Economic Adjustment Programme for Greece' to reduce public expenditure, a restructuring which had a significant negative impact on hospitality SMEs (Kathimerini, 2012). In midst of the financial crisis, social tourism received some new attention in the political discourse as a means to support the elderly in a period of pension and benefit cuts, and to boost the domestic market (Ministry of Economy, Development and Tourism, 2013). As a result of this attention, social tourism programmes were relaunched in 2013. In 2014, however, EOT's 'Tourism for All' programme (including its sub-programme for seniors), the major social tourism programme in Greece, stopped running, and since then, it has been inoperative.

Currently, social tourism activity in Greece is mainly organised and run by the Manpower Employment Organisation (OAED), and the Agricultural Insurance Organisation (OGA), while also encompassing smaller private providers (e.g. 'KEMEN Hellas'). Public programmes operate all year round, and offer partially-funded holidays, through coupons (subsidies), to different target groups, namely, employed and unemployed individuals, seniors, people with disabilities, and their protected family members (the carers of people with a 67% disability or more, are also eligible to participate). The eligibility criteria vary to some extent, depending on each programme's provider; however, what all public programmes have in common, is that their beneficiaries live on low-income, and have not participated in any other social tourism programme over the past year. With regards to accommodation suppliers, all types of hotels, rooms to let, and campsites around Greece with a valid 'Special Operation Logo' are eligible to participate.

The duration of holidays, the number of beneficiaries, and the amount of subsidy differ amongst providers, and also fluctuate (to a larger or lesser extent) every year, depending on the amount of financing they receive from the Greek Government. For example, two recent publicly funded programmes run by OAED (2015) and OGA (2016), offered holidays (up to five overnight stays) to 150,000 and 55,000 beneficiaries, respectively. With regards to the amount of subsidy, in the period 2013–2014, for instance, OAED's programme subsidised 7,00 € to 22,00 € per person per night, depending on the chosen type of accommodation, whereas EOT's 2013–2014 programme subsidised 6,00 € to 14,00 € per person per night. For the same period OAED's budget was 10,000,000 € (OAED, 2013), and EOT's 5,000,000 € (EOT, 2013). Although not constant, the minimum contribution of beneficiaries has been low (e.g. 1,00 € per person per night for EOT's programmes).

Recently, a new public initiative has been incorporated into the existing social tourism programmes run by OAED and OGA, aiming to boost the economic activity in the North Aegean Islands (e.g. Lesvos, Kos, Chios, Samos, and Leros), which have experienced a drop in tourist demand due to the refugee crisis in the region. These sub-programmes cover the full accommodation costs (up to ten overnight stays) for beneficiaries who will choose to one of the above destinations for their holidays (OAED, 2016).

This paper argues that social tourism can significantly contribute to the mitigation of some of the limitations of mass tourism, as it has the following attributes: (a) it runs in the off-season period, aiming to extend the tourist flow beyond the summer-season; (b) its services are largely provided by SMEs, the backbone of the Greek economy, which has been almost crashed during the recession; and (c) its positive effects on people's physical and mental health benefit a large proportion of the population who has suffered the multiple consequence of this recession. Hence, social tourism addresses some of the major socioeconomic problems of Greece, such as unemployment, acute poverty, the shrinking of SMEs, and social exclusion (e.g. IOBE, 2016; Ministry of Economy, Development and Tourism, 2013), and in doing so, it is also consistent with the principles of sustainable development.

Policy recommendations

Considering the multiple benefits that emanate from the practice of social tourism, it is argued here that social tourism can play the role of the stabiliser of the Greek tourism system. It can create demand during the off-season period, enabling the continuity of tourist flow beyond the summer-season. In order to achieve this, some changes in the planning, implementation, and funding of social tourism are imperative. Social tourism programmes in Greece, run for a prolonged period of time throughout the year, offering beneficiaries the choice to go on holiday at any time during this period (e.g. between 15 October 2015 and 30 September 2016 for OAED's programme). The result of such a 'loose' implementation period is that social tourism practice fails to be specifically concentrated in the off-season period in Greece (November–April), when is mostly needed. Furthermore, Greece's budgetary limitations and new fiscal reforms, threaten the feasibility of publicly funded programmes in the long-term (Kathimerini, 2016 November).

Arguably, the implementation period of social tourism programmes is necessary to be shortened. It is acknowledged that it is not feasible for social tourism to cover the whole six-month off-season period. Therefore, it is suggested that holiday-breaks should be offered only during the shoulder-season, thus extending the overall tourist season. This suggestion has three possible options, namely, to implement social tourism programmes: (a) at the end of the tourist season; (b) before the beginning of the new season; and (c) both at the end of the tourist season and before the beginning of the new season. Although there is no right or wrong choice, implementing social tourism programmes before the beginning of the new season (e.g. March and April) combines several advantages. Firstly, it concerns the two months prior to the start of the summer-season, thus, aiming to contribute to an earlier start of the new tourist season; secondly, the tourism demand in this period is stronger than the respective demand recorded during November and December, therefore, an increase in overnight stays and other consumption will have a more sizable impact on the economy; and thirdly, the risk of non participation is lower, given that the weather in spring is better, and the Greek National Holiday and Easter Holidays (including a two-week school holiday) fall into this period of time.

Although the total number of social tourists in Greece per year is unknown due to the lack of available data, considering the latest press releases from OAED (2015) and OGA (2016), it is estimated that there are approximately 180,000 beneficiaries each year.[1] In reality, the total number is expected to be larger, given that there is also provision of social tourism holidays from private providers. With a maximum of five overnight stays that the publicly funded programmes offer each beneficiary, the total of overnight stays is approximately 900,000 per year. With a relatively even distribution of overnight stays in March and April, social tourism can contribute approximately 450,000 overnight stays to each of these months, strengthening a currently weak period of tourist demand,

and contributing to a smoother transition from the off-season to the summer-season (Table 1).

In other words, it is argued here, that social tourism can act as an automatic stabiliser within the Greek tourism sector, and the local economies at destinations, helping increased consumption over a longer period of time, and, thus keeping local markets alive and in relative equilibrium (Keynes, 1997 [1936]). Similarly to traditional automatic stabilisers, such as unemployment benefits, which cannot fully compensate for the consumption that is generated through the spending of the employed segment of the population, social tourism by no means can counterbalance the spending of international tourists. Its economic contribution is modest, yet, sustainable, given that low-income groups and their need for tourism participation will continue to exist in the future. In addition, the more stable economic activity that will be gradually generated, as a result of social tourism's demand, could offer a fruitful ground for the development of further economic activity both related and unrelated to tourism. For instance, a more stable economic environment within a region that suffers from seasonality could potentially strengthen the development of alternative forms of tourism (e.g. agritourism), and the creation of other SMEs which will respond to the increased demand for local products (e.g. agricultural), or lead to new international market demand. Indicative of these effects are findings from a recent study by Eusébio et al. (2016) on the impact of social tourism programmes at destinations in Portugal. The study found that social tourism for seniors had positive effects on local communities, not only in relation to the mitigation of seasonality, but also with regards to diversification and revitalisation of local economies.

On the other hand, budgetary limitations cannot be ignored since they comprise a major obstacle for the sustainability of social tourism programmes in Greece. Therefore, estimating the costs and potential returns of social tourism practice is imperative both for policy-makers and the public opinion (e.g. tax-payers). The 'Social Tourism 2015–2016' programme, for instance, had a 10 million € budget (OAED, 2015). Although there are no available data in order to estimate the tax revenue for the State as a result

Table 1. Overnight stays of residents and non-residents in hotels, similar establishments, and tourist campsites, by month (2015) – Potential contribution of social tourism.

Month	Hotels and similar establishments		Tourist campsites			Including social tourists	
	Residents	Non-residents	Residents	Non-residents	Total	New total	Increase %
January	681,159	371,789	250	1,164	1,054,362		
February	742,976	373,277	181	828	1,117,262		
March	865,899	589,159	263	995	1,456,316	1,906,316	30.9
April	1,012,549	1,970,666	2,571	8,391	2,994,177	3.444,177	15.0
May	1,125,634	7,093,175	51,578	60,798	8,331,185		
June	1,124,979	10,822,252	93,961	107,743	12,148,665		
July	1,575,011	13,398,566	214,188	232,015	15,419,780		
August	2,512,015	13,410,856	338,118	266,574	16,527,563		
September	1,170,481	10,240,428	62,730	104,473	11,578,112		
October	845,364	4,432,517	1,580	8,802	5,288,263		
November	684,955	500,960	462	1,446	1,187,823		
December	860,251	367,195	361	1,027	1,228,834		

Note: The category 'Residents' possibly includes social tourists, too. This possibility, however, is highly unlikely to have a strong influence on the figures for March and April, given the trend amongst Greeks to travel mainly during mid-summer (ELSTAT, 2016b).
Source: Adapted from ELSTAT (2016b).

of the economic activity that social tourism has generated, data from similar initiatives abroad, show that it is significant. For example, the Spanish State has received (or recovered) 1.32 € in taxes (e.g. VAT) and other savings (e.g. social protection) for every euro invested in the pilot transnational programme 'Europe Senior Tourism' (European Commission, 2010b). Furthermore, given that in Greece the maximum amount of contribution, per beneficiary per night, is 5 € on average, there is a maximum spending of 3.75 million € for accommodation only (OAED, 2015). With additional per person spending on food, drink and activities to be taken into consideration, it could be argued that with a modest public investment, social tourism generates important economic benefits both for the State and the local markets at destinations. In addition, social tourism helps SMEs (the cornerstone of the Greek private sector, which has almost collapsed as a result of the deep economic recession) to survive at the present, and grow in the future. Yet, a major obstacle in the case of Greece is the extent to which the State can implement a functional tax-system.

Given these potential financial returns, it is suggested that a smoother transition from the off-season to the summer-season could be further assisted by an increase in the number of social tourists. Admittedly, the current number of beneficiaries is particularly low given the socioeconomic characteristics of a large proportion of the Greek population. According to the latest available data, in 2015, 35.7% of the population lived at risk of poverty or social exclusion, and 40.7% of the population experienced material deprivation, including the inability to afford a one-week holiday every year (ELSTAT, 2016a). This means that the beneficiaries of publicly funded social tourism holidays account only for the 3.5% of the population living at risk of poverty or social exclusion, and the 3.1% of those experiencing material deprivation.[2] Hence, if Greece wants to utilise social tourism's potential more fully, it needs to increase (at least gradually) the proportion of individual beneficiaries, through additional public or public-private funding. This increase will result in a more sizable contribution to the country's sustainable development goals, such as further reduction of social exclusion, increase of citizens' (both social tourists' and residents' of host-communities) well-being and quality of life, and will further boost the economic activity and employment rates at destinations. Recent research on INATEL and IMSERSO social tourism programmes in Portugal and Spain, respectively, confirms this potential (see Cisneros-Martínez et al., 2017; Eusébio et al., 2013, 2016). Although it is acknowledged that any increase in tourist flow can potentially have negative environmental effects, this is highly unlikely in the case of social tourism for several reasons. For instance, social tourism largely concerns a domestic form of tourism, which adopts environmentally conscious practices (e.g. use of public transport and more eco-friendly accommodation), and is practiced in neglected areas during the off-season (see Baumgartner, 2011; Cisneros-Martínez et al., 2017). Moreover, the volume of social tourists is significantly lower than the volume of 'mainstream' tourists, and it can always be controlled by the public authorities which are the main providers of social tourism programmes.

These recommendations do not imply that social tourism can stand alone. Hence, it must not be misconstrued as a single strategy that operates independently from mass tourism, other forms of tourism, and other sectors. In contrast, it is suggested here that social tourism should be fully integrated both into the tourism, and the wider socioeconomic, system of Greece. With regards to the former, tourism can act as the means that ensures the continuum of tourist flow for a prolonged period of time, thus playing the role of a successive and complementary strategic tool to mass and alternative forms of tourism. Within this equation, social tourism can assist the sustainability of the tourism sector as a whole. For instance, after the end of the off-season period, the tourism system will continue to operate as usual, mainly through international arrivals, but it will do so under different conditions than before. SMEs will be in a more favourable position in terms of revenues and liquidity ability, which will allow them to better compete. In turn, this will result in a more fair competition between large businesses and SMEs, thus, reducing the inequality of opportunities and prospects within the private sector, and contributing to the sustainability of SMEs. Finally, within a more stable and fair economic environment, any unsustainable practices, which are often encouraged by financial strain (e.g. tax-evasion) are more likely to fade (Bramwell, 1998).

Conclusions

This paper has discussed the potential of social tourism to contribute to sustainability outcomes for host-communities at destinations, by focusing on the case of Greece. After highlighting the strong conceptual linkages between social tourism and sustainable development, and the limitations of Greek tourism as it has been traditionally operated, the paper identified that social tourism can address some of these limitations, and in doing so, it does offer potential to mitigate against the major socioeconomic problems (e.g. unemployment, acute poverty, and social exclusion) which have overwhelmed Greece since the beginning of the recent financial crisis. This potential of social tourism has not been realised as yet however, mainly due to the lack of proper planning and implementation of social tourism programmes, and the vulnerability of social tourism to cuts in public expenditure. Therefore, the paper has proposed some tangible policy recommendations with the aim to unleash this potential. Specifically, we suggested that social tourism programmes should run exclusively during the shoulder-season in order to contribute to the extension of the tourist season, and by doing so, stabilise the socioeconomic situation in tourism communities at destinations. In addition, we stressed the need for an increase in the number of social tourists through public or public-private partnership funding. If this were to happen, social tourism could address both tourism-specific and wider societal issues (e.g. Eusébio et al., 2016), all of which impede the country's sustainable development. This contribution may be modest, but it is expected to bring about positive, long-term effects.

On the other hand, it must be acknowledged that the conceptual character of this paper imposes certain limitations. Given the absence of empirical data, the economic benefits of social tourism have been presented as simple approximations. Similarly, any wider social benefits have not been empirically explored. Hence, there is a need for empirical studies that will measure the specific economic benefits of social tourism for host-communities, and will explore the wider social impact of these benefits. Furthermore, in spite of the very diverse range of social tourism programmes and systems in Europe (McCabe et al., 2011), an attempt to compare findings from different cultural contexts would have possibly strengthened our policy recommendations. Finally, the focus of this paper on Greece's socioeconomic environment has resulted in de-emphasising the environmental pillar of sustainable development. Although the environmental pillar maybe the less sustainable in social tourism in relation to the economic and social pillars, future studies on the potential environmental effects of social tourism programmes on host communities are needed to help us better understand the relationship between social tourism and all three pillars of sustainable development.

Notes

1. OAED offered 150,000 beneficiaries the chance to go on a holiday between 15 October 2015 and 30 September 2016. OGA's programme runs between 23 July 2016 and 8 May 2017 and has 55,000 beneficiaries. In order to estimate how many beneficiaries in total went on a holiday in one year, we have set the time-frame of a year in line with the time-frame of OAED's programme (October–September). Then we added the total number of OAED's beneficiaries to the number of OGA's beneficiaries who went on a holiday until September 2016. Given that the latter figure is unknown, and considering the travel trend of Greeks to go on holidays mainly during the summer (ELSTAT, 2016a), it was assumed that the majority of OGA's beneficiaries went on holidays between July and September. The exact number of this majority is also unknown, and the choice of 30,000 beneficiaries serves as a rather conservative proxy to the true figure.
2. Estimations according to the 2011 Population – Housing Census revision of 20 March 2014 (ELSTAT, 2014).

Disclosure statement

No potential conflict of interest was reported by the authors.

ORCID

Scott McCabe ⓘ http://orcid.org/0000-0002-9807-9321

References

Baumgartner, C. (2011). Social tourism and sustainability. In S. McCabe, L. Minnaert, & A. Diekmann (Eds.), *Social tourism in Europe: Theory and practice* (pp. 166–177). Bristol: Channel View.

Bramwell, B. (1998). Selecting policy instruments for sustainable tourism. In W. F. Theobald (Ed.), *Global tourism* (pp. 361–379). Oxford: Butterworth-Heinemann.

Bramwell, B., & Lane, B. (1993). Sustainable tourism: An evolving global approach. *Journal of Sustainable Tourism, 1*(1), 1–5.

Buhalis, D. (2001). Tourism in Greece: Strategic analysis and challenges. *Current Issues in Tourism, 4*(5), 440–480.

Chi, C. G., Cai, R., & Li, Y. (2017). Factors influencing residents' subjective well-being at World Heritage Sites. *Tourism Management, 63*, 209–222.

Cisneros-Martínez, J. D., McCabe, S., & Fernández-Morales, A. (2017). The contribution of social tourism to sustainable tourism: A case study of seasonally adjusted programmes in Spain. *Journal of Sustainable Tourism.* doi:10.1080/09669582.2017.1319844

Cole, S. (2006). Information and empowerment: The keys to achieving sustainable tourism. *Journal of Sustainable Tourism, 14*(6), 629–644.

Despotaki, G., Tsartas, P., & Doumi, M. (2015, October). *Social tourism as a breath of the crisis in Greece: Research on social tourism institutions.* Paper presented at the 1st International Conference on Experiential Tourism, Santorini, Greece.

EESC. (2006). *Opinion of the European Economic and Social Committee on Social tourism in Europe.* Retrieved from http://eur-lex.europa.eu/legal-content/EN/TXT/?uri=CELEX%3A52006IE1155

ELSTAT. (2014). *2011 Population and housing Census - Housing Census revision of 20/3/2014*. Retrieved from http://www.statistics.gr/en/statistics/-/publication/SAM03/-

ELSTAT. (2016a). *Greece in figures: October-December 2016*. Retrieved from http://www.statistics.gr/en/greece-in-figures

ELSTAT. (2016b). *Arrivals and nights spent in hotels, similar establishments and tourist campsites: 2015*. Retrieved from http://www.statistics.gr/en/statistics/-/publication/STO12/2015

EOT. (2013). *Πρόγραμμα Τουρισμός για Όλους 2013–2014* [Tourism for All 2013–2014 Programme]. Retrieved from http://www.koinonikostourismos.gr/pdfs/tourgiaolous2013_2014.pdf

European Commission. (2010a). *Calypso study on social tourism: Greece*. Retrieved from http://ec.europa.eu/DocsRoom/documents/6936?locale=en

European Commission. (2010b). *Calypso study on social tourism: Spain*. Retrieved from http://ec.europa.eu/DocsRoom/documents/6948?locale=en

European Commission. (2014). *Annual Tourism Report 2013 - Greece*. Retrieved from http://ec.europa.eu/DocsRoom/documents/5957/

Eurostat. (2013). *Sustainable development indicators*. Retrieved from http://ec.europa.eu/eurostat/web/sdi/indicators

Eusébio, C., Carneiro, M. J., Kastenholz, E., & Alvelos, H. (2013). The economic impact of health tourism programmes. In A. Matias, P. Nijkamp, & M. Sarmento (Eds.), *Quantitative methods in tourism economics* (pp. 153–173). Heidelberg: Physica.

Eusébio, C., Carneiro, M. J., Kastenholz, E., & Alvelos, H. (2016). The impact of social tourism for seniors on the economic development of tourism destinations. *European Journal of Tourism Research, 12*, 5–24.

Goulet, D. (1992). Development: Creator and destroyer of values. *World Development, 20*(3), 467–475.

Higgins-Desbiolles, F. (2006). More than an "industry": The forgotten power of tourism as a social force. *Tourism Management, 27*(6), 1192–1208.

IOBE. (2016). *The Greek economy* (No. 82). Retrieved from http://iobe.gr/docs/economy/en/ECO_O4_15_REP_ENG.pdf

Kakoudakis, K. I., McCabe, S., & Story, V. (2017). Social tourism and self-efficacy: Exploring links between tourism participation, job-seeking and unemployment. *Annals of Tourism Research, 65*, 108–121.

Kathimerini. (2012, February 25). *«Λουκέτο» σε ξενοδοχεία φέρνει η κατάργηση των προγραμμάτων κοινωνικού τουρισμού* [The abolishment of social tourism programmes brings hotel shutdowns]. Retrieved from http://www.kathimerini.gr/451515/article/oikonomia/epixeirhseis/loyketo-se-3enodoxeia-fernei-h-katarghsh-twn-programmatwn-koinwnikoy-toyrismoy

Kathimerini. (2016, August 27). *«Κλειδί» για την ανάπτυξη της ελληνικής οικονομίας οι ξένες επενδύσεις* [Foreign investments, 'key' for the development of the Greek economy]. Retrieved from http://www.kathimerini.gr/872418/article/oikonomia/ellhnikh-oikonomia/kleidi-gia-thn-anapty3h-ths-ellhnikhs-oikonomias-oi-3enes-ependyseis

Kathimerini. (2016, November 5). *Νέα κριτήρια στη χορήγηση κοινωνικών επιδομάτων και έλεγχος των παροχών* [New criteria for the administration of social benefits and provisions' control]. Retrieved from http://www.kathimerini.gr/882206/article/oikonomia/ellhnikh-oikonomia/nea-krithria-sth-xorhghsh-koinwnikwn-epidomatwn-kai-elegxos-twn-paroxwn

Keynes, J. M. (1997). *The general theory of employment, interest, and money*. New York, NY: Prometheus Books (Originally published 1936).

Kim, K., Uysal, M., & Sirgy, M. J. (2013). How does tourism in a community impact the quality of life of community residents? *Tourism Management, 36*, 527–540.

Liu, Z. (2003). Sustainable tourism development: A critique. *Journal of Sustainable Tourism, 11*(6), 459–475.

Lu, J., & Nepal, S. K. (2009). Sustainable tourism research: An analysis of papers published in the Journal of Sustainable Tourism. *Journal of Sustainable Tourism, 17*(1), 5–16.

McCabe, S., & Johnson, S. (2013). The happiness factor in tourism: Subjective well-being and social tourism. *Annals of Tourism Research, 41*, 42–65.

McCabe, S., Minnaert, L., & Diekmann, A. (Eds). (2011). *Social tourism in Europe: Theory and practice*. Bristol: Channel View.

McDonald, J. R. (2009). Complexity science: An alternative worldview for understanding sustainable tourism development. *Journal of Sustainable Tourism, 17*(4), 455–471.

Ministry of Economy, Development and Tourism. (2013). *Ομιλία Υπουργού Τουρισμού Όλγας Κεφαλογιάννη: Τουρισμός και τρίτη ηλικία* [Speech of the Minister of Tourism Olga Kefalogianni: Tourism and the third-age]. Retrieved from http://www.mintour.gov.gr/el/PressRoom/PressReleases/?EntityID=7fd170d8-fdfa-4e21-b0ba-7a9618cfe6c0

Ministry of Economy, Development and Tourism. (2014). *Κατευθύνσεις Εθνικής Αναπτυξιακής Στρατηγικής 2014–2020 στον τομέα του τουρισμού* [Direction plan of the National Development Strategy 2014–2020 for the tourism sector]. Retrieved from http://www.mintour.gov.gr/el/Investments/Espa/Administrationservice/NewProgram20142020/

Minnaert, L., Maitland, R., & Miller, G. (2009). Tourism and social policy: The value of social tourism. *Annals of Tourism Research, 36*(2), 316–334.

Morgan, N., Pritchard, A., & Sedgley, D. (2015). Social tourism and well-being in later life. *Annals of Tourism Research, 52*, 1–15.

OAED. (2013). *Δημόσια πρόσκληση 02/2013 για το πρόγραμμα επιδότησης διακοπών με επιταγή κοινωνικού τουρισμού έτους 2013–2014* [Public call 02/2013 for the 2013–2014 programme of subsidised holidays through social tourism coupons]. Retrieved from http://www.koinonikostourismos.gr/pdfs/dhmosiaproskloaed2013.pdf

OAED. (2015). *Δημόσια πρόσκληση 06/2015 για το πρόγραμμα επιδότησης διακοπών με επιταγή κοινωνικού τουρισμού έτους 2015–2016* [Public call 06/2015 for the 2015–2016 programme of subsidised holidays through social tourism coupons]. Retrieved from http://www.oaed.gr/documents/10195/1224065/dimproskl16.pdf/849a7ffc-43f5-49b8-bf26-6789da7a2801

OAED. (2016). *Συμπληρωματική δημόσια πρόσκληση 5/2016 για το πρόγραμμα επιδότησης διακοπών με επιταγή κοινωνικού τουρισμού έτους 2015–2016* [Supplementary public call 5/2016 for the 2015–2016 programme of subsidised holidays through social tourism coupons]. Retrieved from http://www.oaed.gr/koinonikos-tourismos-2015-2016

OECD. (2001). *The DAC Guidelines: Strategies for sustainable development.* Retrieved from http://www.oecd.org/dac/environment-development/2669958.pdf

OECD. (2011). *How's life? Measuring well-being.* Paris: OECD Publishing.

OGA. (2016). *Πρόγραμμα Κοινωνικού Τουρισμού 2016* [Social Tourism Programme 2016]. Retrieved from http://www.oga.gr/index.php?pg=estia1

Ryan, C. (2002). Equity, management, power sharing and sustainability - issues of the 'new tourism'. *Tourism Management, 23*(1), 17–26.

SETE. (2016). *Ελληνικός Τουρισμός: Εξελίξεις – Προοπτικές* [Greek Tourism: Developments – Prospects]. Retrieved from http://sete.gr/media/5444/periodiki-meleti-ellinikos tourismos_teyhos-1.pdf

Sharpley, R. (2000). Tourism and sustainable development: Exploring the theoretical divide. *Journal of Sustainable Tourism, 8*(1), 1–19.

UN. (2010). *Sustainable development: From Brundtland to Rio 2012.* New York, NY: United Nations.

WCED. (1987). *Our Common Future.* Retrieved from http://www.un-documents.net/our-common-future.pdf

Woo, E., Kim, H., & Uysal, M. (2015). Life satisfaction and support for tourism development. *Annals of Tourism Research, 50*, 84–97.

'Not on my vacation': service encounters between able-bodied and disabled consumers – the case of high-contact service

Anat Tchetchik, Victoria Eichhorn* and Avital Biran*

ABSTRACT

The effect of service encounters on customer satisfaction in high-contact services is gaining wider recognition among academics and practitioners alike. In this exploratory study, we aimed at gaining insight into a specific service encounter, namely, between able-bodied and disabled customers. While many studies have addressed the experiences of disabled persons in a plethora of situations, including tourism and leisure settings, the service experience of able-bodied customers sharing a service environment with disabled customers has been overlooked. Using a simple choice experiment, we showed that when given textual information about the expected presence of disabled guests in advance, two-thirds of the able-bodied study participants would be willing to stay in a hotel with a minor to moderate probability of being co-present with disabled guests. However, when a visual aid was provided together with the textual information, this figure decreased significantly. Implications and potential recommendations for policy-makers and hotel managers are presented, as are suggestions for future research. Such research is imperative if better inclusion of disabled tourists is to be achieved, with the resulting realisation of the immense underexploited economic potential of disabled tourism.

RESUMEN

El efecto de los encuentros de servicio en la satisfacción del cliente en servicios de contacto elevado está ganando un mayor reconocimiento tanto entre académicos como entre los prestadores del servicio. En este estudio exploratorio, perseguimos comprender mejor un encuentro de servicio específico: aquel entre clientes con plenas capacidades y clientes con discapacidad. Mientras que muchos estudios han analizado las experiencias de las personas con discapacidad en una plétora de situaciones, incluyendo los marcos de turismo y ocio, se ha ignorado la experiencia de servicio de clientes con plenas capacidades compartiendo un entorno de servicio con clientes con discapacidad. Utilizando un experimento de elección simple mostramos que, dada una información textual previa sobre la presencia esperada de huéspedes con discapacidad, dos tercios de los participantes en el estudio con plenas capacidades

---

*These authors contributed equally to this work.

estarían dispuestos a alojarse en un hotel con un probabilidad de baja a moderada de presencia de huéspedes con discapacidad. Sin embargo, cuando se proporcionaba una ayuda visual junto con la información textual, esta cifra descendía significativamente. Se presentan las implicaciones y recomendaciones potenciales para los responsables de formulación de políticas y los gestores de hotel así como sugerencias para investigaciones futuras. Tal investigación es imperativa si se busca una mayor inclusión de los turistas discapacitados con el resultado de la realización del inmenso potencial económico subexplotado del turismo para discapacitados.

RÉSUMÉ
L'effet des expériences de service sur la satisfaction de la clientèle dans les services de proximité est de plus en plus reconnu dans le secteur académique et professionnel. L'objectif de cette étude exploratoire était d'approfondir notre compréhension des expériences de service spécifique, à savoir celles entre clients actifs et les handicapés. Alors que de nombreuses études ont abordé les expériences des personnes handicapées dans diverses situations, y compris le tourisme et les loisirs, l'expérience de service des clients actifs partageant un environnement de service avec des clients handicapés n'a pas eu l'attention nécessaire. À l'aide d'une expérimentation basée sur un simple choix, nous avons montré que si les informations écrites relatives à la présence des personnes handicapées étaient disponibles à l'avance, les deux tiers des participants actifs à l'étude seraient disposés à partager l'hôtel avec les handicapés, avec une petite ou moyenne probabilité. Cependant, lorsqu'une aide visuelle et les informations textuelles ont été fournies, ce chiffre a diminué de façon significative. Cette étude identifie des implications et émet des recommandations potentielles destinées aux décideurs des politiques et les directeurs d'hôtels, tout en suggérant de nouvelles pistes de recherche. De nouvelles recherches sont nécessaires pour parvenir à une meilleure stratégie inclusive tenant compte des touristes handicapés, ce qui permettrait l'exploitation effective de l'immense potentiel économique sous-exploité du tourisme favorable aux personnes handicapées.

摘要
无论在学术界还是业界，在高接触型服务中服务接触对消费者满意度的影响收到越来越多的重视。在这个探索性的研究中，我们的目的是探索一个特别的服务接触——在健全和残疾消费者之间的服务接触。许多研究已经在很多情景下探索过残疾人士的体验，包括旅游和休闲情境，但是，健全消费者和残疾消费者一同共享一个服务环境的情景下的服务体验却被忽视了。研究使用一个简单选择实验发现，当提前告知健全游客有关残疾客人会出现的背景信息，三分之一的健全参与者愿意住在一个有小至中等可能性会和残疾可以共处的酒店中。然而，当背景信息以及视觉辅助都提供了，这一个人群数量显著地减少。文章为政策制定者以及酒店管理者提供了指导以及可能的建议，以及为未来的研究方向提供了建议。如果想让残疾游客更好地被容纳，这方面的研究势在必行，并且可以实现残疾旅游开发不足的巨大的经济潜质。

Introduction

Customer satisfaction – or dissatisfaction – with customer services, especially those categorised as high-contact services, is often affected by the behaviour and characteristics of other customers who are part of the service experience (Pranter & Martin, 1991). This satisfaction/dissatisfaction will then be reflected in the customers' appraisal of the delivered service. The findings of Grove, Fisk, and Dorsch (1998), for example, have highlighted that 'other customers' are responsible for the smallest and largest proportion of satisfying and dissatisfying situations, respectively, when the effect of 'other customers' is compared to the effects of service setting, service providers and service performance. A classic example of the significant effect of 'other customers' may be found in air travel. David Wilson of *World Travel* magazine has identified nine types of flyer which travellers would prefer to avoid as 'seat neighbours', for example, intoxicated or loud/talkative passengers. To minimise the occurrence of undesirable customer encounters, airlines such as Air Baltic and KLM allow passengers to state their preferences for a seat partner when making a booking. KLM operates this service by offering passengers the option of sharing parts of their Facebook, Google+ or LinkedIn profiles, of seeing the profiles of other passengers on their prospective flight, and hence of choosing seats next to someone they regard as good in-flight company (Computerworlduk, 2014). Other examples in the service sector include the grouping of people accompanied by pets or of families with children or the smokers in smoking lounges. Such 'grouping' efforts reflect the growing attention of many service companies to customer compatibility, a process that aims at attracting homogeneous groups of customers to their service environment. According to Wu (2007), customer compatibility is positively related to the homogeneity of the customer group in terms of similar preferences, sought benefits, attitudes, past experiences and physical characteristics.

Physical characteristics, in particular, are attributes that can lead to a type of service encounter between disabled and able-bodied consumers that could affect the service experience of the involved customers. Moral and ethical considerations and national and international legislation notwithstanding, such service encounters cannot always be controlled or monitored by service providers. In the U.S.A., for example, the Americans with Disabilities Act (ADA) of 1990 was introduced with the aim of enhancing the independence, participation and equality of opportunities for people with disabilities. This legislation has led to an increase in the proportion of people with disabilities fully participating in many activities and consuming services previously not accessible to them (Packer, McKercher, & Yau, 2007) and hence to an increased likelihood of encounters between disabled and able-bodied service consumers. Nonetheless, it seems that many service providers and particularly tourism and hospitality businesses try to avoid complying with this legislation. As a result, despite the fact that the ADA explicitly requires modification of hospitality policies and procedures to ensure that guests with disabilities are provided with a level of services equal to that offered to able-bodied guests, inconsistencies in the accessibility of accommodation, restaurants, theatres and museums remain a problem (National Council on Disability, 2007). Therefore, the proportion of people with disabilities fully participating in mainstream tourism activities is only slowly increasing, in contrast to the situation in the job market (Packer et al., 2007).

It seems that the reluctance to assure physical access is not the only barrier preventing service providers from providing adequate services for disabled consumers. Other barriers include the negative attitudes of the service employees in the travel and hospitality industry towards travellers with physical disabilities (Daruwalla & Darcy, 2005; McKercher, Packer, Yau, & Lam, 2003). These attitudes are reflected, for example, by lack of empathy, unconstructive behaviour, rude behaviour and even small insensitive gesture or body language (Kim & Lehto, 2012), or in the tendency not to address the person with a disability directly but rather to speak to his/her companion instead (Poria, Reichel, & Brandt, 2011). Based on our interpretation of the literature, such attitudes may reflect one or both of the following assumptions: (1) the industry holds negative attitudes towards disabled people and even to the extent that the provider complies with legal regulations by ensuring physical access, the behaviour and practices of the employees remain unchanged. (2) Management silently assumes that the presence of disabled customers will constitute a source of customer incompatibility which will ultimately affect the way the guests assess the service encounter. Indeed, contrary to Daruwalla and Darcy (2005), Kim and Lehto (2012) and others' arguments regarding hotels' employees negative attitudes, there are some few examples from the industry that indicate an intrinsic desire by service providers to create an environment of customer compatibility, for example, Hotel Weisseespitze in Austria which explicitly market the presence of disabled customers in the co-presence of motorcycle riders, or Hotel Atlantic in Prague who promotes itself as 'the Barrier-Free Hotel'. However, the resulting perceptions of these service experiences have not been studied in the context of the service sector, in general, or that of tourism and hotel vacations, in particular. Furthermore, previous studies on disability have been focused on the points of view of disabled customers (Lovelock, 2010; Darcy & Pegg, 2011) or of the service providers (Daruwalla & Darcy, 2005) but not on those of able-bodied customers.

This study thus aims to gain preliminary insights into the service experience of able-bodied hotel guests who share a service environment with disabled guests. Understanding service encounters between able-bodied and disabled vacationers from the point of view of able-bodied customers is important for two main reasons. First, it is necessary to either refute or confirm the notion that the presence of disabled guests presents a source of customer incompatibility which negatively affects the customer experiences of able-bodied guests. Confirming this assumption entails that the service experience of disabled guests is also negatively affected since even if able-bodied guests do not express their feelings verbally, much of what we communicate is nonverbal including facial and body gestures. Moreover, when verbal communication contradicts nonverbal ones', people usually believe the nonverbal one and rely on nonverbal behaviour to judge another's attitudes and feelings (Burgoon, Guerrero, & Floyd, 2016).

Second, if such an assumption is indeed found to have valid grounds, it will be necessary to understand the extent, nature, characteristics and factors affecting this phenomenon as a prerequisite to designing measures to enhance customer satisfaction for all guests. This will ultimately result in a greater inclusion of disabled people in tourism, travel and hospitality activities, since real inclusive holiday experiences cannot be achieved without improving the able bodied-disabled nexus in service encounters. This notion is supported by past studies emphasising that, apart from the importance of the physical environment, a key element in triggering feelings of exclusion is the social environment, as reflected by the reaction of non-disabled people to people with disabilities (Eichhorn, Miller, & Tribe, 2013; Poria et al., 2011). Yet, changing the social environment may prove a difficult task: despite some degree of public acceptance, able-bodied individuals evidently still exhibit a plethora of negative attitudes and emotions towards disabilities. These include anxiety and fear, discomfort, embarrassment, pity and active avoidance from interactions with individuals with have visible disabilities (Albrecht, Walker, & Levy, 1982;

Bruce, Harman, & Baker, 2000; Fichten, Amsel, Robillard, Sabourin, & Wright, 1997). These attitudes and emotions do not exist in a vacuum but are shaped by the specific context and situation. The situational context hypothesis posits that the situational context affects attitudes toward disabled people and that both the type of disability and the social context affect the formation of attitudes. Grand, Bernier, and Strohmer (1982) were the first to explore the attitudes of able-bodied people towards people with various disabilities (cerebral palsy, epilepsy, amputation and blindness) in relation to social situations, such as work, dating and marriage. In the situational context, the highest acceptance levels were found in work situations, and in the context of disabilities the lowest acceptance level was evident for cerebral palsy. Importantly, no previous study has explored this hypothesis in the service environment.

The current study thus aims to address this knowledge gap. Drawing on the situational context hypothesis, we address a social situation that differs from those studied in the past, namely, the service encounter in a service environment that is characterised by high-contact and high-involvement experiences. We focus on the service experience of able-bodied customers in the presence of visibly disabled customers during a hotel vacation. Notably, due to its nature, a hotel vacation offers a unique situational context: While the situations examined in previous studies (e.g. Grand et al., 1982; McCaughey, 2010) are located on a scale of commitment (work, dating and marriage) and the interpersonal interactions involved in such commitments, hotel vacations cannot be easily located on such a scale: the shared hotel service environment does not inherently involve – or carry an obligation for – interpersonal interactions between guests, but it does entail sharing public spaces, such as swimming pools and saunas, which forces physical intimacy between strangers. In this study, we employed a quantitative approach to conduct a preliminary investigation to gain an initial understanding of this type of encounter from the perspective of able-bodied consumers. We focussed on the indirect effect of such encounters – namely, the effect that stems from disabled guests being part of the service environment – rather than the direct effect that takes place during interpersonal encounters. Using the stated-preference technique, we provided the respondents with a choice task and instructed them to choose one hotel out of a set of alternative hotels. The respondents were provided with information that is not always available when looking for a hotel stay, i.e. information about the expected presence of disabled guests in the hotel.[1] The information was given in one of two forms, textual alone or textual plus a visual aid (photographs). Consumers' choices including the effect of the two types of information on their choices were analysed by using an econometric modelling.

Methodology

Research methodology

To address the research objective, a discrete-choice framework with a stated-preferences approach was adopted. Participants were asked to complete a questionnaire that included choosing a three-night hotel vacation on the basis of the three options presented in Figure 1. In contrast to real-life situations, participants were informed in advance about the expected presence of disabled guests during their vacation and were requested to make their choice based on this information. A simple choice task was designed in which each respondent had to choose one out of three hotels. Other parts of the questionnaire included questions regarding vacationing habits, attitudes towards disability, including previous familiarity with disability (family, work, etc.), and the emotions towards people with disability. From a previously identified range of emotions towards people with disabilities (Albrecht et al., 1982; Corrigan, Green, Lundin, Kubiak, & Penn, 2001; Corrigan et al., 2002; Fiske, Cuddy, Glick, & Xu, 2002; Hirschberger, Florian, & Mikulincer, 2005), the research team and a pilot group chose four emotions that were deemed relevant to this case study, such as, pity, fear, disgust and empathy.

Data treatment

One manipulation was employed, namely, some of the participants were provided only with a textual description of the hotels and the expected presence of disabled guests (Figure 1), whereas the other participants received both the textual description of the hotels and brochure-type photographs for each hotel.

The textual description represented the a-priori probability of being co-present with disabled guests, ranging from nearly zero (Hotel 1) to moderate (Hotel 3). The photographs featured disabled guests in the hotel lobby, dining room and swimming pool. Due to the potential ethical issues relating to the staging and photographing of visibly disabled people, the photographs were taken from the Internet. In selecting the photographs, the emphasis was placed on those focusing on the guests and not on the hotel facilities. The suitability and appropriateness of the pictures was tested by a pilot group drawn from a class of BA students.

To create a trade-off between the presence of disabled guests and other hotel attributes, we fixed the prices such that the price was always highest for Hotel 1 (not accessible for disabled guests) and lowest for Hotel 3 (most accessible). The pricing reflected the assumption, based on the above-mentioned previous studies indicating that able-bodied

Suppose you decided to spend a long weekend (3 nights) in the city of Eilat in a chain-affiliated 4-star hotel. While searching for a hotel, you are faced with the three alternatives presented below. All stated prices are for a family (couple + 1, 2, or 3 children) for 3 nights, including breakfast. Assuming all other attributes are equal, which hotel would you choose?

HOTEL 1	HOTEL 2	HOTEL 3
Not accessible for disabled guests	Partially accessible for disabled guests, i.e., 30% of rooms are accessible, as are the lobby and the dining areas. You could expect a minor presence of physically disabled guests during your stay	Highly accessible for disabled guests, i.e., 60% of rooms are accessible, as are the lobby, dining areas, and all spa facilities including the swimming pool. You could expect a moderate presence of physically disabled guests during your stay
Price: $ 1167	Price: $ 1055	Price: $ 944

Figure 1. Choice Task – Textual description alone.

people hold negative attitudes towards disabilities, that people have no a-priori positive preference to be in the presence of disabled people and would not be willing to pay extra to be in their presence *ceteris paribus*. Nonetheless, the price differences were not very high, the maximum difference being 20%, which, considering the luxury good nature of a hotel vacation, is not excessive.

Study population and sampling methodology

The study population comprised adult (above 18 years) able-bodied individuals. Given the sensitivity of the topic, snowball sampling was used. The survey was conducted online so as to avoid bias towards socially desirable responses due to the presence of an interviewer. The number of valid observations used for the data analysis was 242 ($n = 81$ with photographs and $n = 161$ with textual description alone). It should be noted that the number of distributed questionnaires with and without photographs was equal. However, some of the respondents who received the questionnaires with the photographs did not return their questionnaires, presumably because they were not willing to participate in the study. Nevertheless, we were able to obtain significant results with the number of responses obtained. Appendix 1 provides the descriptive statistics of the sample participants.

Empirical results

Preliminary results

As can be seen from Figure 2, employing a simple t-test for mean differences, the distribution of the hotel choice of the participants provided with photographs was significantly different from that of the participants provided with textual information alone, with the percentage of people choosing Hotel 1 (most expensive; no accessibility) and Hotel 2 (intermediate price; moderate accessibility) being significantly higher in the 'text + photograph' group.

Econometric results

To better understand participants' decision-making processes, an econometric approach was employed. Since the dependent variable is an ordered categorical variable (disabled guests' co-presence level), an ordered logit with bootstrap standard error model was chosen, and a reduced form model that explains the choice of hotel was regressed. The benefit of using a regression model is that it allows controlling for the effect of other independent variables. Not surprisingly, respondents who have received a textual description plus photographs of the hotels reported significantly higher levels of pity, fear, disgust and empathy than those who received the textual description alone, which indicates a problem of endogeneity, i.e. the mere showing of the photographs evoked stronger emotions. This issue was addressed by running auxiliary regressions, namely, each emotion variable was regressed over the variable indicating the research method, and the residuals of these regressions were plugged into the main regression model to establish the level of each emotion that was not explained by the use of visual demonstration. Table 1 presents

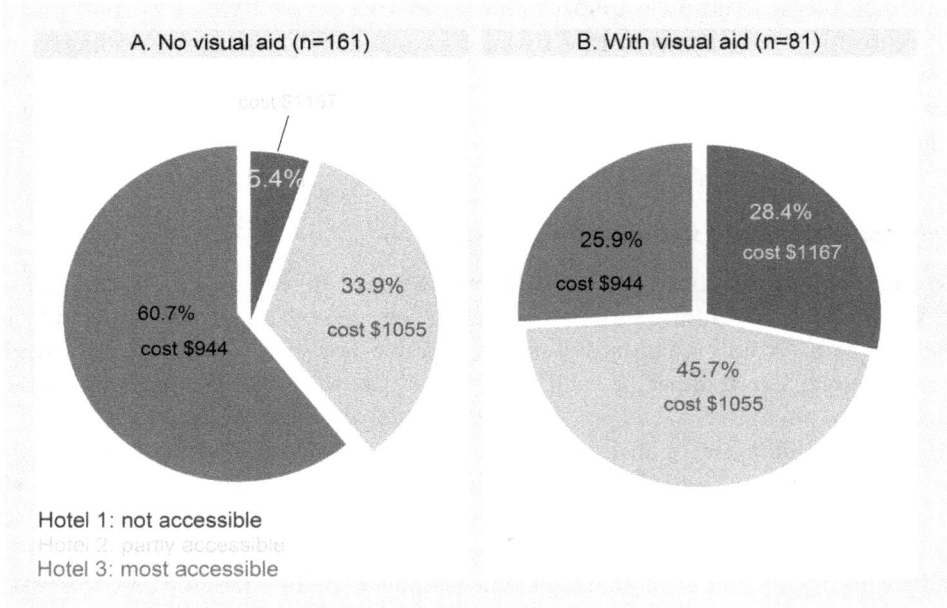

Figure 2. Distribution of hotel choice by type of choice.

Note: t tests indicated that the differences between the percentage of people choosing each type of hotel for the participants presented with the visual aid + text and those presented with the text alone are significant at $p < .05$ in all cases.

the calculated marginal effects, i.e. how much a unit change in an independent variable
changes the probability to choose Hotel 1.

The results indicate that the photographs had a significantly positive effect on the probability of choosing a hotel with no presence of disabled people. Among the four emotion
variables, *pity*, *fear* and *disgust* had positive and significant effects on the preference for
Hotel 1 over the other two hotels. Secularity also had a significantly positive effect on
the preference to spend the vacation in a 'non-accessible' hotel. In contrast, the probability
of choosing Hotel 1 was negatively and significantly affected by an upbringing on a
kibbutz and by friendships with disabled people. Finally, as was to be expected, income
had a positive, but non-significant, effect on the probability of choosing Hotel 1. This
finding may be attributed to the fact that the differences in the prices of the three
hotels were relatively small. Notably, although our experimental design used a trade-off
between vacation price and the presence of guests with visible disabilities, it was not
our intention to estimate willingness to pay per se but rather to reveal the magnitude of
preferences.

Discussion

The disabled population is an important consumer segment. In the U.S., the number of
people with disabilities is anticipated to reach 100 million by the year 2030 (Ozturk,
Yayli, & Yesiltas, 2008). In Europe, a report prepared by the European Commission indicated that more than half of Europeans with disabilities travelled in 2012 and 2013

Table 1. Ordered logit regression with bootstrapped standard errors.

Variable	Description	Marginal effects after ordered logit[a] y = Pr (choice = 1)
Visual aid	=1 if participant was shown photographs	0.161*** (0.047)[b]
Family	=1 if participant is familiar with disabled persons in his/her family	−0.015 (0.033)
Friendship	=1 if participant is familiar with disabled persons via friendship	−0.052** (0.023)
School	=1 if participant is familiar with disabled persons from school	−0.024 (0.035)
Work	=1 if participant is familiar with disabled persons in his/her work	0.019 (0.043)
Pity residual	Residual of level of pity[c] regressed over the research method dummy	0.026** (0.012)
Fear residual	Residual of level of fear[c]	0.028* (0.017)
Disgust residual	Residual of level of disgust[c]	0.027* (0.015)
Empathy residual	Residual of level of empathy[c]	−0.011 (0.011)
Children	=1 if participant has children under 10 years old	−0.005 (0.016)
Kibbutz	=1 if participant was raised in a collective community (kibbutz) in Israel	−0.059** (0.021)
Gender	=1 if participant is a male	0.035 (0.026)
Religiosity	=1 if participant defines him/herself as secular	0.046** (0.021)
Income	a categorical variable from 1 (lowest) to 6 (highest)	0.005 (0.008)

Note: $N = 242$, Pseudo $R^2 = 0.2$.
*$p < .10$, **$p < .05$, ***$p < .01$.
[a]dy/dx for dummy variables refers to a discrete change of the variable from 0 to 1.
[b]Standard errors are given in parentheses.
[c]Scale of 1 (very little) to 5 (greatly).

(European Commission, 2013). The report predicts that by 2020 the general demand for accessibility will account for 154.6 million, representing an annual increase of 1.2%. Moreover, it was also found that: (1) 70% of the market demanding accessibility have both the financial and the physical capabilities to travel and (2) disabled tourists tend to take longer holiday breaks than the average and spend more per day than able-bodied tourists (Horgan-Jones & Ringaert, 2004). Accordingly, it is suggested that if airlines, hotels, restaurants, museums and other tourist facilities will be fully accessible, this market can generate potential revenues of €88.6 billion by 2025 (Bowtell, 2015). Notwithstanding, since travelling with a disability is more than an access issue (Yau, McKercher, & Packer, 2004), there is another barrier to overcome before the full market potential can be reached. This barrier refers to the intangible components of the service experience, which according to Grady and Ohlin (2009) can matter just as the tangible ones. The literature provides evidence regarding the attitudes of the service staff towards travellers with physical disabilities including apathy, lack of empathy or flexibility, unconstructive behaviour, rude behaviour and even small insensitive gesture or body language (Kim & Lehto, 2012). This negative approach of hotel's service staff will not change until the management of the hotel proactively adopts positive attitudes towards guests with physical disabilities.

Yet, the literature on customer compatibility and its effect on the service experience and the growing importance of creating superior customer experiences suggests that hotels managers may be reluctant to adopt positive attitudes towards guests with physical disabilities due to concerns of customer incompatibility. In that sense, our results highlight a positive message: Even when photos of physically disabled hotel guests were provided, only one-quarter of the respondents were not willing to stay in hotels where the service space will be co-shared with disabled guests. This figure dropped to 5.4% when no visual aid was provided. This negates to a large extent hotels' managements concerns of customers' incompatibility.

Secondly, 80% of the respondents in our snowball based sample are less than 40 years old, of which 75% are between 18 and 28 years old. These cohorts were born around the 1990s to a cultural and regulatory environment that condemns discrimination against people with physical disabilities. This is reflected in the findings that younger respondents demonstrated significantly less fear disgust and pity towards people with physical disabilities compared to respondents above 40 years old (see Appendix 2A). We believe that this evidence evolved since the ADA became law, and prove that people's attitudes toward persons with disabilities are going through changes (Takeda & Card, 2002).

Thirdly, our findings suggest that for some respondents disability evokes emotions such pity, fear and disgust, which reduces the willingness to stay in an accessible hotel. These results are in line with findings that feelings and emotions are linked to most aspects of consumer behaviour and that they constitute a principal component in the perceptions of service experiences (Arnold & Reynolds, 2009; Mattila & Enz, 2002). At the same time, the current research reveals factors that increase the willingness to be co-present with disabled people while on holiday (which constitutes an additional situation-specific information). Specifically, being raised in a kibbutz or having personal friendships with people with a disability correspond with significantly lower levels of emotions like pity or disgust (see Appendix 2B). Thus, familiarity and sensibilisation from early childhood with people with a disability (as is cultivated on kibbutzim in Israel or by contact with disabled friends and family members) are found to be key drivers for inclusion that should be further encouraged.

These findings support the literature that emphasises the need for more education and training, involving face-to-face contact with disabled people from early childhood, to encourage positive attitudes and acceptance (Paez & Arendt, 2014; Schitko & Simpson, 2012). Improved acceptance levels are particularly important for the formation of personal attitudes in the face of societal attitudes. While the former refers to beliefs and opinions that individuals possess with regard to certain individuals, the latter relates to widespread attitudes held by society at large, influenced by civil and legal rights (Daruwalla & Darcy, 2005). In the context of a hotel vacation, positive societal attitudes might be in place (i.e. the belief that it is wrong to discriminate against people with a disability), but deeply rooted negative personal attitudes, that ultimately prevent positive and enriching encounters between disabled and able-bodied guests, may still be very much present, reinforcing strong sentiments of exclusion faced by individuals with a disability.

Policy and managerial implications

The exploratory nature of the current study confines the ability to draw profound policy implications. Nevertheless, several key practical implications can be identified. First, with the aim to improve the service experience for all guests, it could be argued that policy in this area should be strengthened to allow for a greater co-presence. Thus, with more disabled people participating in a hotel vacation, abled-body guests will gradually stop experience customer incompatibility as it will become a natural part of societal life. This is supported by our finding that familiarity with disability positively affects the willingness to be co-present with disabled hotel guests. Yet, it appears that the strengthening of legal acts alone is insufficient as other obstacles exist. To start with, the private ownership structure of most of the tourism infrastructure represents a major impediment to the removal of barriers (Rains, 2008). Private owners usually regard the laws protecting the rights of disabled people as representing an additional cost category (Imrie & Kumar, 1998) or as legal risks of law suits to be managed (Rains, 2008). In this respect, legal acts generate reluctance on the side of private business resulting in misperceptions of individuals with disabilities as non-lucrative customers (Rains, 2008). This is supported by Imrie and Kumar (1998) stating that the needs of impaired individuals are often perceived as negligible with minimum standards only poorly implemented (Stumbo & Pegg, 2005), which prevents an overall change in attitudes of service providers as well as of other customers.

Thus, to create an environment where everyone enjoys and participates equally would require the implementation of Universal Design, a human-centred framework to design places, information, communication and policy in a way that benefits the broadest range of individuals in the amplest range of situations without the need for specialised adaptations (Veitch & Shaw, 2004; Horgan-Jones & Ringaert, 2001; Rains, 2008). That way, disability is recognised as part of human diversity (Darcy & Harris, 2003) which accommodates disabled individuals without stigmatising them (Brown, Kaplan, & Quaderer, 1999). In line with this, it has been suggested that wider community-oriented inclusion frameworks are needed to improve the co-participation of people with disabilities in community life in general (Fujimoto, Rentschler, Le, Edwards, & Härtel, 2014). Such programmes should be tailored to the establishment of common interest activities between diverse individuals, allowing for the creation of a shared social identity of individuals regardless of their level of ability. This shared social identity forms the basis for a greater acceptance of and willingness for co-presence. Yet, little is known about common interest activities in the specific context of tourism.

However, until fundamentals change, such as the full implementation of Universal Design has been taken place, our findings carry an important message with managerial implications for hotels. As only a relatively small share of our abled-body respondents chose to stay in hotels with no presence of disabled guests, it dismisses to some extent any concerns hotels managers' may have regarding customers' incompatibility. Taken together with the potentially high revenue of the market demanding accessibility, hotel managers will be better off not only complying to physical accessibility regulations but also to genuinely welcome guests with disabilities. This should be accompanied by training hotel's employees to accommodate disabled guests as part of the customer experience. Our results suggest that familiarity with disabled people increases the willingness for co-presence. Thus, training hotel's employees to accommodate disabled guest could be achieved by hiring disabled hotel staff. This will also contribute to the goal of including more people with disabilities in the job market. Notably, Houtenville and Kalargyrou (2015, p. 168) argue that the 'leisure and hospitality businesses are less likely than goods-producing industries to report that the nature of the work is such that it cannot be effectively performed by people with disabilities'. Moreover, Laabs (1994), who examined the Chicago Marriott Hotel, found that turnover rate dropped to 32% per year after it started employing people with disabilities. These findings further reinforce the benefits of hiring disabled hotel staff.

Finally, on a marketing perspective, it has been argued that for tourism to be inclusive, specifically tailored inclusive promotional material is needed (Darcy & Pegg, 2011). Nonetheless, the current empirical findings indicate a paradoxical situation, as the visualisation of people with disabilities evoked more intense emotions such as pity, fear and disgust, which are apparently undesirable in the context of a hotel vacation and reduce the willingness of able-bodied vacationers to stay in an accessible hotel. This does not necessarily mean that promotional material that includes images of people with physical disabilities should be avoided. Notably, the visual aid employed in our study was drawn from the contemporary presentation of disability in the media. This presentation 'is pretty much the same as it has always been: clichéd, stereotyped and archetypal' (Darke, 2004, p. 100) and includes narratives focusing on charity, health and personal-tragedy stories. Thus, a representation of disability in the media is needed before images, language and terminology related to disability can be effectively included in hotels' promotional material. The need for representing disability in the media was stressed by the disabled community and organisations and is gaining growing public and academic attention in the recent decade (Müller, Klijn, & Van Zoonen, 2012; Beacom, French, & Kendall, 2016). Yet, it appears that the media have been slow to respond despite the massive coverage of the Paralympics games in 2012 and 2014, hoping to be a real chance to change wider public perceptions about disability.

Limitations and suggestions for future research

This study provides preliminary insights into the willingness of able-bodied people to be co-present with individuals with disability during high-contact service encounters, particularly in the context of a hotel vacation. As the implications of such co-presence are much broader, it is suggested that future studies should use larger, representative and cross-culture samples with additional experiments. Field studies and experimental studies employing neurobiological methods, such as fMRI and eye tracking devices, are also warranted, since they could provide further knowledge about the nature of these service encounters and perhaps even suggest means to reduce undesirable emotions evoked during these encounters. Further manipulations in relation to the visual demonstration can be employed, for example, examining the impact of advertising material of hotels that portray disabled guests using alternative narratives as in the case of several brands who employ disabled presenters to communicate messages of empowerment (e.g. Nike, which employed the Paralympic gold medallist sprinter, Oscar Pistorius, or Volvo who employed Bethany Hamilton, a professional surfer who lost an arm after a shark attack, in its campaign). Such studies could help identify specific marketing interventions that could promote a greater inclusion.

Future studies should also include respondents envisaging a holiday vacation with a disabled friend or family. According to the UN (2011), together with their families, the number of people directly affected by disability is approximately 2 billion people – a third of the global population. Many guests with physical disabilities are accompanied by family members and/or friends, which are not necessarily disabled. This group might have a different point of view and preferences. In that vein, it seems that the expanding literature which addresses the inclusion of people with disabilities in tourism has overlooked the challenges faced by guests who accompany disabled family members or friends to a hotel stay or other tourism/leisure activities. Future research ought to pay attention to the preferences, barriers, challenges and experiences sought by these visitors.

Another issue relates to the motivations for going on a vacation. Crompton (1979) identified several motivations: 'escape from a perceived mundane environment, exploration and evaluation of self, relaxation, prestige, regression, enhancement of kinship relationships, facilitation of social interaction', but also novelty and education. It would be of interest to explore how each of these motivations interacts with co-presence service encounters in the hotel environment.

Finally, other service domains should be studied beyond the hotel vacation, as such studies would provide a further understanding of the unique characteristics of different service environments and experiences. Adding more service domains to future studies could also assist in providing greater insights into the type of common interest activities that are needed to improve co-participation structures for both tourism and the everyday life in general. This may help to develop an understanding *by everyone* that vacation sites are places for relaxation and stress avoidance, benefiting all vacationers as well as of the players in the industry.

Note

1. While some hotels indirectly provide information on how much they welcome disabled guests in their marketing communication, e.g. http://www.sagetraveling.com/london-wheelchair-accessible-hotelshttp://www.sagetraveling.com/, others do not.

Disclosure statement

No potential conflict of interest was reported by the authors.

References

Albrecht, G. L., Walker, V. G., & Levy, J. A. (1982). Social distance from the stigmatized: A test of two theories. *Social Science & Medicine, 16*(14), 1319–1327.

Arnold, M. J., & Reynolds, K. E. (2009). Affect and retail shopping behavior: Understanding the role of mood regulation and regulatory focus. *Journal of Retailing, 85*(3), 308–320.

Beacom, A., French, L., & Kendall, S. (2016). Reframing impairment? Continuity and change in media representations of disability through the paralympic games. *International Journal of Sport Communication, 9*(1), 42–62.

Bowtell, J. (2015). Assessing the value and market attractiveness of the accessible tourism industry in Europe: A focus on major travel and leisure companies. *Journal of Tourism Futures, 1*(3), 203–222.

Brown, T. J., Kaplan, R., & Quaderer, G. (1999). Beyond accessibility: Preferences for natural areas. *Therapeutic Recreation Journal, 33*, 209–221.

Bruce, A. J., Harman, M. J., & Baker, N. A. (2000). Anticipated social contact with persons in wheelchairs: Age and gender differences. *Advances in Psychology Research, 1*, 219–228.

Burgoon, J. K., Guerrero, L. K., & Floyd, K. (2016). *Nonverbal communication.* Routledge.

Computerworlduk. (2014). *KLM uses social media to allow passengers to choose their ideal seat neighbors*. Retrieved from http://blogs.computerworlduk.com/news/it-business/3509905/klm-uses-social-media-to-allow-passengers-to-choose-their-ideal-seat-neighbours/

Corrigan, P. W., Green, A., Lundin, R., Kubiak, M. A., & Penn, D. L. (2001). Familiarity with and social distance from people who have serious mental illness. *Psychiatric Services, 52*(7), 953–958.

Corrigan, P. W., Rowan, D., Green, A., Lundin, R., River, P., Uphoff-Wasowski, K., ... Kubiak, M. A. (2002). Challenging two mental illness stigmas: Personal responsibility and dangerousness. *Schizophrenia Bulletin, 28*(2), 293.

Crompton, J. L. (1979). Motivations for pleasure vacation. *Annals of Tourism Research, 6*(4), 408–424.

Darcy, S., & Harris, R. (2003). Inclusive and accessible special event planning: An Australian perspective. *Event Management, 8*, 39–47.

Darcy, S., & Pegg, S. (2011). Towards strategic intent: Perceptions of disability service provision amongst hotel accommodation managers. *International Journal of Hospitality Management, 30*, 468–476.

Darke, P. (2004). The changing face of representations of disability in the media. In J. Swain, S. French, C. Barnes, & C. Thomas (Eds.), *Disabling barriers-enabling environments* (2nd ed., pp. 100–105). London: Sage Publications.

Daruwalla, P., & Darcy, S. (2005). Personal and societal attitudes to disability. *Annals of Tourism Research, 32*, 549–570.

Eichhorn, V., Miller, G., & Tribe, J. (2013). Tourism: A site of resistance strategies of individuals with a disability. *Annals of Tourism Research, 43*, 578–600.

European Commission. (2013). *Economic impact and travel patterns of accessible tourism in Europe* (Final Report, Service Contract SI2. ACPROCE052481700). European Commission, DG Enterprise and Industry.

Fichten, C. S., Amsel, R., Robillard, K., Sabourin, S., & Wright, J. (1997). Personality, attentional focus, and novelty effects: Reactions to peers with disabilities. *Rehabilitation Psychology, 42*(3), 209–230.

Fiske, S. T., Cuddy, A. J., Glick, P., & Xu, J. (2002). A model of (often mixed) stereotype content: Competence and warmth respectively follow from perceived status and competition. *Journal of Personality and Social Psychology, 82*(6), 878–902.

Forgas, J. P. (1995). Mood and judgment: The affect infusion model (AIM). *Psychological Bulletin, 117*(1), 39–66.

Fujimoto, Y., Rentschler, R., Le, H., Edwards, D., & Härtel, C. E. J. (2014). Lessons learned from community organizations: Inclusion of people with disabilities and others. *British Journal of Management, 25*(3), 518–537.

Grady, J., & Ohlin, J. (2009). Equal access to hospitality services for guests with mobility impairments under the Americans with disabilities act: Implications for the hospitality industry. *International Journal of Hospitality Management, 28*, 161–169.

Grand, S. A., Bernier, J. E., & Strohmer, D. C. (1982). Attitudes toward disabled persons as a function of social context and specific disability. *Rehabilitation Psychology, 27*(3), 165–174.

Grove, S. J., Fisk, R. P., & Dorsch, M. J. (1998). Assessing the theatrical components of the service encounter: A cluster analysis examination. *The Service Industries Journal, 18*(3), 116–134.

Hirschberger, G., Florian, V., & Mikulincer, M. (2005). Fear and compassion: A terror management analysis of emotional reactions to physical disability. *Rehabilitation Psychology, 50*(3), 246–257.

Horgan-Jones, M., & Ringaert, L. (2001, October 14–16). *Accessible tourism in Manitoba*. TTRA – Travel and Tourism Research Association. Niagara Falls, Canada.

Houtenville, A., & Kalargyrou, V. (2015). Employers' perspectives about employing people with disabilities: A comparative study across industries. *Cornell Hospitality Quarterly, 56*(2), 168–179.

Imrie, R., & Kumar, M. (1998). Focusing on disability and access in the built environment. *Disability & Society, 13*, 357–374.

Kim, S. E., & Lehto, X. Y. (2012). The voice of tourists with mobility disabilities: Insights from online customer complaint websites. *International Journal of Contemporary Hospitality Management, 24*(3), 451–476.

Laabs, J. J. (1994). Individuals with disabilities augment Marriott work force. *Personnel Journal (USA), 73*(9), 46–51.

Lovelock, B. A. (2010). Planes, trains and wheelchairs in the bush: Attitudes of people with mobility-disabilities to enhanced motorised access in remote natural settings. *Tourism Management, 31*(3), 357–366.

Mattila, A. S., & Enz, C. A. (2002). The role of emotions in service encounters. *Journal of Service Research, 4*(4), 268–277.

McCaughey, T. (2010). *Individual and situational factors associated with social barriers for persons with mobility impairment (Doctoral dissertation).* University of Illinois at Urbana-Champaign.

McKercher, B., Packer, T., Yau, M., & Lam, P. (2003). Travel agents as facilitators or inhibitors of travel: Perceptions of people with disabilities. *Tourism Management, 24*(4), 465–474.

Müller, F., Klijn, M., & Van Zoonen, L. (2012). Disability, prejudice and reality TV: Challenging disablism through media representations. *Telecommunications Journal of Australia, 62*(2), 28.1–28.13.

National Council on Disability, USA. (2007). *Implementation of the Americans with Disabilities Act: challenges, best practices, and new opportunities for success.* Retrieved from http://www.ncd.gov/publications/2007/July262007

Ozturk, Y., Yayli, A., & Yesiltas, M. (2008). Is the Turkish tourism industry ready for a disabled customer's market? The views of hotel and travel agency managers. *Tourism Management, 29* (2), 382–389.

Packer, T. L., McKercher, B., & Yau, M. (2007). Understanding the complex interplay between tourism, disability and environmental contexts. *Disability & Rehabilitation, 29*(4), 281–292.

Paez, P., & Arendt, S. W. (2014). Managers' attitudes towards people with disabilities in the hospitality industry. *International Journal of Hospitality and Tourism Administration, 15,* 172–190.

Poria, Y., Reichel, A., & Brandt, Y. (2011). Dimensions of hotel experience of people with disabilities: An exploratory study. *International Journal of Contemporary Hospitality Management, 23* (5), 571–591.

Pranter, C. A., & Martin, C. L. (1991). Compatibility management: Roles in service performers. *Journal of Services Marketing, 5*(2), 43–53.

Rains, S. (2008). Culture in the further development of universal design. *Design for All, 3,* 18–34.

Schitko, D., & Simpson, K. (2012). Hospitality staff attitudes to guests with impaired mobility: The potential of education as an agent of attitudinal change. *Asia Pacific Journal of Tourism Research, 17*(3), 326–337.

Stumbo, N. J., & Pegg, S. (2005). Travelers and tourists with disabilities: A matter of priorities and loyalties. *Tourism Review International, 8,* 195–209.

Takeda, K., & Card, J. A. (2002). Us tour operators and travel agencies: Barriers encountered when providing package tours to people who have difficulty walking. *Journal of Travel & Tourism Marketing, 12*(1), 47–61.

UN. (2011). *World report on disability.* Retrieved from http://www.who.int/disabilities/world_report/2011/report.pdf

Veitch, C. & Shaw, G. (2004, January). Access and tourism: A widening agenda. In British Tourist Authority (Ed.). *Insights - Tourism Intelligence Papers, A-121.*

World Travel. (2012). *Who are the worst people to sit next to on a plane?* Retrieved from http://travel.ninemsn.com.au/world/804100/who-are-the-worst-people-to-sit-next-to-on-a-plane

Wu, C. H. J. (2007). The impact of customer-to-customer interaction and customer homogeneity on customer satisfaction in tourism service—the service encounter prospective. *Tourism Management, 28*(6), 1518–1528.

Yau, M. K. S., McKercher, B., & Packer, T. L. (2004). Traveling with a disability: More than an access issue. *Annals of Tourism Research, 31*(4), 946–960.

Appendices

Appendix 1. Descriptive statistics of the sample's respondents.

Variable	Description	Mean
Male	=1 if respondent is a male	50%
Single	=1 if respondent does not live with a spouse	37%
Have_Kids	=1 if respondent has children until 18 years old	35%
Kids_04	=1 if respondent has children between the ages of 0-4	14%
Kids_59	=1 if respondent has children between the ages of 5-9	18%
Kids_1014	=1 if respondent has children between the ages of 10-14	16%
Kids_1518	=1 if respondent has children between the ages of 15-18	31%
City	=1 if respondent was born & raised in a city	78%
Kibbutz	= '–' in a collective community	10%
Moshav	= '–' in a cooperative agricultural community	7%
High school	=1 if respondent has high school education	16%
student	=1 if respondent is a student	40%
BA	=1 if respondent has BA degree	22%
MA	=1 if respondent has MA degree	14%
Tertiary education	=1 if respondent has professional non-academic education	4%
Secular	=1 if respondent is secular	81%
Traditional	=1 if respondent is traditional	12%
Religious	=1 if respondent is religious	7%
Income_0	=1 if respondent has no work income at the time of the survey	9%
Income_1	=1 if respondent's income is much below the average	30%
Income_2	=1 if respondent's income is below the average	24%
Income_3	=1 if respondent's income equals the average	13%
Income_4	=1 if respondent's income is above the average	18%
Income_5	=1 if respondent's income is much above the average	7%
Age_1	=1 if age is between 18 and 28	60%
Age_2	=1 if age is between 29 and 40	21%
Age_3	=1 if age is between 41 and 52	10%
Age_4	=1 if age is between 53 and 64	5%
Age_5	=1 if age is 65+	3%
Family	=1 if participant is familiar with disabled persons in his/her family	20%
Friendship	=1 if participant is familiar with disabled persons via friendship	16%
School	=1 if participant is familiar with disabled persons from school	9%
Work	=1 if participant is familiar with disabled persons in his/her work	15%

Appendix 2. t tests with unequal variances.

A. Extent of feelings towards disabled people by age category:

Feel fear

Group	Obs	Mean	Std. Err.	Std. Dev.	[95% conf. Interval]	
Age1|age2	201	1.567	0.023	0.333	1.521	1.614
Age1|age2|age3	38	2.079	0.061	0.376	1.955	2.202
combined	239	1.649	0.025	0.388	1.599	1.698
diff		−0.511	0.065		−0.643	−0.380
diff = mean(0) − mean(1)		$t = -7.8223$				
Ho: diff = 0	Satterthwaite's degrees of freedom = 48.573					
Ha: diff < 0	Ha: diff ! = 0	Ha: diff > 0				
Pr(T < t) = 0.000	Pr(|T| > |t|) = 0.000	Pr(T > t) = 1.000				

Feel Pity

Group	Obs	Mean	Std. Err.	Std. Dev.	[95% conf. Interval]	
Age1|age2	201	3.453	0.039	0.555	3.376	3.530
Age1|age2|age3	38	3.814	0.106	0.654	3.599	4.029
combined	239	3.510	0.038	0.585	3.436	3.585
diff		-0.361	0.113		-0.588	-0.133

diff = mean(0) − mean(1) $t = -3.190$
Ho: diff = 0   Satterthwaite's degrees of freedom = 47.59
Ha: diff < 0   Ha: diff ! = 0   Ha: diff > 0
Pr(T < t) = 0.001   Pr(|T| > |t|) = 0.003   Pr(T > t) = 0.998

Feel disgust

Group	Obs	Mean	Std. Err.	Std. Dev.	[95% conf. Interval]	
Age1|age2	201	1.729	0.025	0.359	1.679	1.779
Age1|age2|age3	38	2.040	0.058	0.360	1.922	2.158
combined	239	1.778	0.024	0.376	1.730	1.826
diff		-0.311	0.064		-0.439	-0.184

diff = mean(0) − mean(1) $t = -4.894$
Ho: diff = 0   Satterthwaite's degrees of freedom = 51.908
Ha: diff < 0   Ha: diff ! = 0   Ha: diff > 0
Pr(T < t) = 0.000   Pr(|T| > |t|) = 0.000   Pr(T > t) = 1.000

B. Extent of feelings towards disabled people by being raised in a Kibbutz:
Feel Pity

Group	Obs	Mean	Std. Err.	Std. Dev.	[95% conf. Interval]	
Kibbutz	22	3.182	0.123	0.578	2.925	3.438
other	217	3.544	0.039	0.577	3.467	3.621
combined	239	3.510	0.038	0.585	3.436	3.585
diff		0.362	0.129		0.096	0.628

diff = mean(0) − mean(1) $t = 2.79$
Ho: diff = 0   Satterthwaite's degrees of freedom = 25.423
Ha: diff < 0   Ha: diff ! = 0   Ha: diff > 0
Pr(T < t) = 0.995   Pr(|T| > |t|) = 0.009   Pr(T > t) = 0.004

Feel disgust

Group	Obs	Mean	Std. Err.	Std. Dev.	[95% conf. Interval]	
Kibbutz	22	1.409	0.073	0.340	1.258	1.560
Other	217	1.816	0.024	0.359	1.768	1.864
Combined	239	1.778	0.024	0.376	1.730	1.826
diff		0.406	0.076		0.249	0.564

diff = mean(0) − mean(1) $t = 5.311$
Ho: diff = 0   Satterthwaite's degrees of freedom = 25.984
Ha: diff < 0   Ha: diff ! = 0   Ha: diff > 0
Pr(T < t) = 1.000   Pr(|T| > |t|) = 0.000   Pr(T > t) = 0.000

The holiday-related predictors of wellbeing in seniors

Marlène Mélon, Stefan Agrigoroaei, Anya Diekmann and Olivier Luminet

ABSTRACT

There has been an increase in research on the relationship between holidays and wellbeing in the last decade. However, only a few studies have investigated this association in seniors and the impact of holiday-related predictors of wellbeing is understudied. The aims of this study were to: 1) compare the profile of senior tourists and senior non-tourists on socio-demographic indicators, health, physical activity, and social relations, 2) compare the profile of senior tourists and senior non-tourists on wellbeing, after adjusting for control variables, and 3) examine the impact of holiday-related predictors (frequency, mean duration, frequency of physical, social, cognitive and relaxing activities, degree of perceived health benefits) on wellbeing of senior tourists, over and above the role of various relevant covariates. A sample of 4130 seniors ($M_{age} = 68.2$ years, $SD = 5.8$, range 60–85) filled out a questionnaire related to the last holiday, daily activities, health, and wellbeing. Results showed that senior tourists were younger, more educated, wealthier, and healthier than senior non-tourists. In addition, the levels of wellbeing were higher in senior tourists compared to senior non-tourists, after adjusting for control variables. Hierarchical regressions analyses revealed that frequent holidays, a greater frequency of social and cognitive activities, as well as the degree of perceived health benefits were associated with higher wellbeing.

RESUMEN

Ha habido un incremento en la investigación sobre la relación entre las vacaciones y el bienestar durante la última década. Sin embargo, sólo unos pocos estudios han investigado esta asociación en los mayores y el impacto de los predictores relacionados con las vacaciones y el bienestar está sin estudiar. Los objetivos de estudio eran: 1) comparar el perfil de los mayores turistas y no-turistas en cuanto a indicadores socio-demográficos, salud, actividad física y relaciones sociales; 2) comparar el perfil de los mayores turistas y no turistas en bienestar, tras ajustar las variables de control; y 3) examinar el impacto de los predictores relacionados con las vacaciones (frecuencia, duración media, frecuencia de las actividades físicas, sociales, cognitivas y de relax, grado de beneficios percibidos en la salud) en el bienestar de los turistas mayores, más allá del rol de varias covariaciones relevantes. Una muestra de 4.130 mayores ($M_{edad} = 68,2$ años, $SD = 5.8$, rango 60–85) completaron un cuestionario relacionado con sus últimas vacaciones, actividades diarias, salud y bienestar. Los resultados mostraron que los turistas mayores eran más jóvenes, con mayor nivel de educación, con mayores ingresos y gozaban de mayor salud que los mayores no-turistas. Además, los niveles de bienestar eran superiores en los turistas mayores comparados con los no-turistas, tras ajustar las variables de control. El análisis de regresiones jerárquicas reveló que las vacaciones frecuentes, una mayor frecuencia de actividades sociales y cognitivas, así como el grado en el de beneficios percibidos para la salud estaban asociados con un nivel de bienestar más elevado.

RÉSUMÉ

La recherche sur la relation entre les vacances et le bien-être a connu un grand développement au cours de la dernière décennie. Cependant, seules quelques études se sont penchées sur cette association parmi les personnes âgées et il n'y a pas eu d'études sur l'impact des prédicteurs de bien-être liés aux vacances. Les objectifs de cette étude étaient: 1) de comparer le profil des touristes seniors et les seniors non-touristes sur les indicateurs sociodémographiques, la santé, l'activité physique et les relations sociales, 2) comparer le profil des touristes seniors et des seniors non-touristes en ce qui concerne le bien-être, après ajustement en faveur des variables de contrôle, et 3) examiner l'impact des facteurs prédictifs liés aux vacances (fréquence, durée moyenne, fréquence des activités physiques, sociales, cognitives et relaxantes, degré des bénéfices pour la santé) sur le bien-être des touristes seniors en tenant compte du rôle de diverses covariables importantes. Un échantillon de 4 130 seniors (Moyenne = 68,2 ans, écart-type = 5,8, intervalle de 60 à 85) a rempli un questionnaire portant sur les dernières vacances, les activités quotidiennes, la santé et le bien-être. Après ajustement tenant compte des variables de contrôle, les résultats ont montré que les touristes seniors étaient plus jeunes, plus éduqués, plus riches et plus en santé que les seniors non-touristes. De plus, les niveaux de bien-être étaient plus élevés parmi les touristes seniors que chez les seniors non-touristes. Les analyses de régressions hiérarchiques ont révélé que des congés fréquents, une plus grande fréquence d'activités sociales et cognitives, ainsi que le degré d'avantages pour la santé étaient associés à un bien-être plus élevé.

摘要

在过去的十年之间，对假期与健康之间关系的研究有所增加。然而，只有少数研究调查了老年人和幸福感的关系，幸福感的假期相关的预测指标没有得到充分的研究。这项研究的目的是：1）比较老年游客和老年非游客在社会人口指标、健康状况、身体活动和社会关系的情况，2）在调整控制变量之后，比较老年游客和老年非游客的幸福感的状况，3）研究假期相关的预测指标（频率，平均持续时间，身体、社交、认知和放松活动的频率，感知健康效益的程度）对老年游客幸福感的影响，以及各种相关协变量的作用。在一个 4,130名老年人的样本中（中位数 = 68.2岁，标准差 = 5.8，范围 60–85）填写了与上次假期、日常活动、健康和幸福感有关的问卷。结果显示，老年游客比老年非游客更年轻，受教育程度更高，更富有，更健康。此外，调整控制变量后，老年游客的幸福感水平高于老年非游客。层次回归分析显示，经常度假，社交和认知活动的频率较高，以及感知健康益处的程度与较高的幸福感有关。

Introduction

For a long time, tourism has been mainly considered an economic activity. However, with the increasing number of studies on social tourism and the association with wellbeing, tourism research has shifted to a more social science approach. Thus, tourism started to be considered a social force (Higgins-Desbiolles, 2006), robustly connected to individuals' wellbeing and health. As argued by Smith and Diekmann (2017), tourism studies have extended over time to other disciplines such as psychology and have become more focused on wellbeing in the last few decades. Taking a holiday[1] contributes to the quality of life as they favor social interaction, personal development and individual identity (Richards, 1999). In that perspective, some authors describe the wish to depart for a holiday as a primary desire essential to the quality of life and as a psychological need to slip away from daily life pressure and/or boredom (Richards, 1999; Urry, 1995).

To date, the links between holidays and wellbeing of people have been demonstrated in literature (Chen & Petrick, 2013). However, there are few studies that have been focused on the potential benefits of holidays on wellbeing in seniors (Balderas-Cejudo, Leeson, & Urdaneta, 2017; Ferrer, Sanz, Ferrandis, McCabe, & Garcia, 2016; Gu, Zhu, Brown, Hoenig, & Zeng, 2016). Many tourism studies investigated the profile of the senior market (e.g. Alén, Losada, & de Carlos, 2017) but little is known about the links between the factors associated with holiday experiences and wellbeing in senior tourists.

This study contributes to the tourism literature by providing a psychological perspective on the relationship between holidays and wellbeing in a large sample of seniors. The first approach will consist in comparing the profiles of senior tourists and senior non-tourists on various factors such as socio-demographic indicators, health, general physical activity, and social relations. Second, the association between taking holidays and wellbeing will be examined after adjusting for control variables. In addition, this study responds to a gap in knowledge by exploring the unique contributions of various holiday-related factors, such as type of activities, duration, frequency and perceived benefits, on wellbeing of senior tourists, over and above the role of relevant covariates.

Below, we provide a review of the literature on the links between holidays-related factors and wellbeing of seniors and a description of objectives of this study, following by a description of the method used and data analyses performed. We then present findings from our analyses in relation to research questions, discuss these findings and provide a short conclusion on key findings of this study.

Literature review and objectives

Research has shown that holidays contribute positively to health and wellbeing of people (Diekmann & McCabe, 2016; Wei & Milman, 2002). Multiple studies have documented the positive association between holidays and wellbeing (e.g. De Bloom et al., 2010; Gilbert & Abdullah, 2004; Wei & Milman, 2002), life satisfaction (e.g. Sirgy, Kruger, Lee, & Yu, 2011), quality of life (e.g. Dolnicar, Yanamandram, & Cliff, 2012; Kim, Woo, & Uysal, 2015; Richards, 1999; Sirgy, 2010), positive emotions (e.g. Mitas, Yarnal, Adams, & Ram, 2012) and happiness (e.g. Nawijn, 2011). According to Minnaert and Schapmans (2009), tourism participation can be considered as a form of social intervention tool. Indeed, as pointed out by several studies, holidays have the potential to bring benefits to disadvantaged and even excluded individuals' wellbeing such as the senior citizens (e.g. Dann, 2001; McCabe & Johnson, 2013; Minnaert & Schapmans, 2009). However, despite an increasing number of studies conducted in recent years (Ahn & Janke, 2011; Ferrer et al., 2016; Gu et al., 2016; Hunter-Jones & Blackburn, 2007; Jia et al., 2016; Kim et al., 2015; Moal–Ulvoas & Taylor, 2014; Morgan, Pritchard, & Sedgley, 2015; Nimrod, 2008; Nimrod & Rotem, 2010, 2012; Staats & Pierfelice, 2003; Toepoel, 2013; Wei & Milman, 2002), the relationships between holidays and seniors' wellbeing is still unclear (Balderas-Cejudo et al., 2017; Chen & Petrick, 2013; Gu et al., 2016). Specifically, the existing literature is characterized by a large, heterogeneous spectrum of operationalizations of wellbeing such as psychological wellbeing (Wei & Milman, 2002), life satisfaction (Ferrer et al., 2016), quality of life (Kim et al., 2015), happiness (Staats & Pierfelice, 2003), mood state (Jia et al., 2016), affects (Staats & Pierfelice, 2003) or self-rated health (Gu et al., 2016). In psychology, wellbeing has widely been conceptualized as including two main components (Diener, Oishi, & Lucas, 2003): a positive evaluation of one's life (i.e. life satisfaction, the cognitive component) and everyday positive feelings and moods (i.e. happiness, the affective component). However, in the previous studies, these two components of wellbeing were not systematically taken into account. Another limitation of the majority of the existent studies is that they did not focus on the holiday-related factors specifically associated with wellbeing of senior tourists. Indeed, only a limited number of studies have examined the specific role of different holiday-related activities (e.g. physical, social, cognitive, relaxing) for wellbeing in senior citizens (Kim et al., 2015).

Some scholars did explore the possible mechanisms that could explain the potential benefits of holidays on seniors' wellbeing. Although the psychological dimension of wellbeing was not considered, Gu et al. (2016) recently suggested possible mechanisms that can explain the potential benefits of holidays on seniors' self-rated health: increase of cognitive functioning (e.g. learning experiences, interpersonal communication), frequency of exercises (e.g. walking), social interactions, self-esteem, personal development, and decrease of perceived stress (thanks to escaping of the stressful routine environment). Participation in activities during the holidays represents an important factor associated with the wellbeing in senior tourists (Kim et al., 2015; Nimrod & Rotem, 2010, 2012; Wei & Milman, 2002). The increase of the frequency of activities on holidays has often been reported in studies to explain the potential benefit of a holiday on seniors' wellbeing (Kim et al., 2015). The activity theory (Havighurst, 1961) postulates that 'being involved and maintaining the activities and social interactions of middle age for as long as possible is essential to wellbeing' (Nimrod & Rotem, 2012, p. 380). Wei and Milman (2002) considered the number of activities practiced on holidays (range 0–18) and found a positive association between this factor and wellbeing in a sample of 84 senior tourists. Nimrod and Rotem (2010) highlighted that the more the seniors (age range 50–88) participated in various activities (e.g. educational, physical, cultural, spiritual, etc.) during their last holidays, the more they reported various associated benefits (e.g. general enjoyment, excitement, social bonding, relaxation). Nimrod and Rotem (2010) showed that irrespective of the type of activities (e.g. physical, educational, non-physical) that seniors practiced during holidays, the psychological benefits (such as excitement and relaxation) were present and were relatively similar from one senior to another. In another study, Nimrod and Rotem (2012) showed that holiday-related activities characterized by new experiences were associated with the benefits gained from the tourism experience because they represent opportunities for personal development, which contribute to deepen the sense of meaning in life and then lead to a higher level of wellbeing (Nimrod & Kleiber, 2007).

In addition, tourism experiences provide opportunities to promote social interactions in later life (e.g. Caradec & Petite, 2008; McCabe & Johnson, 2013; Morgan et al., 2015; Toepoel, 2013). Toepoel (2013) showed that seniors (age range 55–75+) who went at least once on holidays over the last 12 months reported lower scores of loneliness compared to seniors who did not go. Many other studies found relationships between social activities practiced on holidays and lower levels of loneliness, the creation of new social interactions and the satisfaction with social contacts in seniors (e.g. Caradec & Petite, 2008; Ferrer et al., 2016; Nimrod & Rotem, 2012; Toepoel, 2013). Moal–Ulvoas and Taylor (2014) highlighted how taking holidays helps senior tourists to improve their relationships with others. These authors suggested that seniors are more open-minded during the holidays, therefore allowing them to better understand others and deepen relationships with them (for example, they communicate and laugh more than in everyday life). Leisure activities such as holiday participation reinforce the feeling of being connected to others (social connectedness, Toepoel, 2013), which in turn contributes to a higher feeling of social inclusion (Ferrer et al., 2016; Gu et al., 2016; McCabe & Johnson, 2013; Minnaert, Maitland, & Miller, 2009; Morgan et al., 2015; Toepoel, 2013).

Together, these different psychosocial mechanisms could help to explain how holidays may be beneficial for seniors' wellbeing. Overall, on holidays, seniors tend to be more relaxed and more available to reflect on their life and to appreciate advanced age (Moal–Ulvoas & Taylor, 2014). Holidays generate a set of positive emotions (Mitas et al., 2012) that allow seniors to reassess their lives in a brighter light and to develop strategies to better cope with the stressors (such as bereavement, illness, and body changes) that occur more often, on average, at their age (Moal–Ulvoas & Taylor, 2014).

Despite the evidence that specific holiday-related activities and experiences have the potential to explain the association with wellbeing, another limitation of the previous studies is that they did not focus on frequency, duration, and perceived health benefits of holidays on wellbeing, although they were identified as important predictors of seniors' health and wellbeing in leisure research (e.g. Chang, Wray, & Lin, 2014; Chen & Petrick, 2014; Chen, Stevinson, Ku, Chang, & Chu, 2012; Williamson, 2016).

Besides studies that have investigated the relationships between holidays and wellbeing, others researchers have examined personal and social factors associated with the participation of seniors citizens in holidays (Ferrer et al., 2016). Some tourism studies have addressed the profile of senior tourists compared to senior non-tourists. For example, there is evidence that both personal (e.g. age, gender, income, health status, lifestyle) and social factors (e.g. social relations) are related to the probability of going on holidays (e.g. Ferrer et al., 2016; Gu et al., 2016). These studies showed that those who had at least one holiday during the last year were more likely to be younger, female, to engage more in leisure activities in everyday life, to have a higher socioeconomic status and better self-rated health compared to senior non-tourists. However, to the best of our knowledge, no studies analysed the difference between both groups in terms of wellbeing, after adjusting for various relevant covariates such as socio-demographic indicators, health status and the frequency of physical and social activities in everyday life.

As part of a broader research project focused on the consequences of senior's holiday participation (BEST[2]), the current study looked more particularly at the relationships between holidays and wellbeing in a large sample ($N = 4130$) of seniors. We systematically explored the relationships between holiday-related factors (frequency, mean duration, frequency of physical, social, cognitive, and relaxing activities, degree of perceived health benefits) and wellbeing in senior tourists, after adjusting for relevant control variables (socio-demographic indicators, self-rated health, general physical activity, and social relations). The specific research questions of this study were the following:

(1) How do the profiles of senior tourists and senior non-tourists differ on socio-demographic indicators, self-rated health, general physical activity, and social relations?
(2) Do senior tourists have higher scores of wellbeing than senior non-tourists, after adjusting for control variables?
(3) What are the holiday-related factors (frequency, mean duration, frequency of physical, social, cognitive and relaxing activities, degree of perceived health benefits) that significantly account for wellbeing in senior tourists, over and above the role of relevant covariates?

With respect to the first research question, in line with previous studies (e.g. Ferrer et al., 2016; Gu et al., 2016), we expected tourists to be younger, female, to frequently participate in leisure and exercise, to have a higher socioeconomic status and higher self-rated health scores compared to senior non-tourists. In addition, we also expected tourists to report higher levels of wellbeing when compared to non-tourists, after adjusting for control variables. To the best of our knowledge, the last research question has not yet been addressed in seniors, using comprehensive wellbeing measures. Thus, our goal was to explore the specific components of holiday experiences and their predictive role for wellbeing in senior tourists.

Method

Participants

A sample of 31,799 French-speaking seniors was selected from the database of a health insurance company in Belgium, a social tourism provider and intermediary (Diekmann & McCabe, 2011). This health insurance company is the largest in Belgium (holding 41.1% of the Belgian market share in 2016). Participants were contacted by email and invited to respond to an online survey about the last holiday, their health and their wellbeing between October and December 2015. The inclusion criteria were: being between 60 and 85 years old, having no functional limitations (based on the Katz score[3]; Katz, Down, Cash, & Grotz, 1970), having no psychiatric and mental disorders, and not having been hospitalized for more than a month during the past year. A total of 5617 seniors filled out the questionnaire (response rate = 18%[4]). As some questionnaires contained more than 5% of missing data (i.e. the threshold considered as consequential; Schafer, 1999), the final sample size was $N = 4130$ (representing a response rate of 13% of the original sample). Compared to the original sample ($N = 31,799$), the final sample ($N = 4130$) showed similar age and gender distributions. In terms of age, in the final sample 65%

were aged 60–69, 30% were aged 70–79, and 5% were aged 80–85. In the original sample, these proportions were 66%, 26% and 5%, respectively. Regarding gender, the final sample comprised of 39.5% of women and 60.5% of men while the original sample had 39.8% and 60.2%, respectively. The higher proportion of men in the final sample could be explained by two observations: the original sample provided by the health insurance company was initially more represented by men (60.2%) and men are more likely than women to respond to online tourism surveys whatever their age (Dolnicar, Laesser, & Matus, 2009).

The descriptive analyses (Table 1) revealed that the mean age of participants was 68.2 years ($SD = 5.8$) and they have attended in school for an average of 12.61 years ($SD = 4.24$). The majority of the respondents were men (60.5%), living with a partner (66.8%), retired (84.9%) and had a net monthly income per household between 1601 and 3000 euros (52.7%). Participants also reported good general health and wellbeing ($M = 3.67$, $SD = .82$; $M = 3.61$, $SD = .71$, respectively, measured both on a five-point scale).

Table 1. Characteristics of respondents.

Variables	n	M	SD
Age (years)	4130	68.2	5.8
Education (years)	4130	12.6	4.24
Gender (% of men)	2500	60.5	
Household composition	3981		
Partner (%)		66.8	
Alone (%)		25.4	
Children (%)		2.7	
Partner and children (%)		4.9	
Living in a community (%)		0.2	
Retired	4130		
Yes (%)		84.9	
Net monthly income per household (in euros)	4051		
Less than 1000 (%)		1.8	
Between 1000 and 1600 (%)		23.7	
Between 1601 and 2200 (%)		26.5	
Between 2201 and 3000 (%)		26.2	
Between 3001 and 5000 (%)		19.6	
Over than 5000 (%)		2.2	
General physical activity[a]	4130	4.09	1.35
Social relations			
Frequency of social contacts[b]	4130	5.41	.93
Quality of social contacts[c]	4130	4.14	.83
Self-rated Health			
General health[d]	4130	3.67	.82
Number of health symptoms (from a list of 11)	4130	2.20	1.75
Functional limitations[e]	4107	2.50	1.26
Self-rated memory failures[f]	4130	2.50	.88
Wellbeing	4130	3.61	.71
Life satisfaction[g]	4130	3.46	.79
Happiness[h]	4130	3.77	.75

Notes: n: sample size, M: mean, SD: standard deviation.
[a] 1 = less than four times a year, 2 = two to three times a year, 3 = one to two times a month, 4 = one to two times a week, 5 = three to six times a week, 6 = over six times a week.
[b] 1 = never, 2 = a few times a year, 3 = one to four times a month, 4 = one to three times a week, 5 = three to six times a week, 6 = each day of the week.
[c] From 1 = not at all satisfied to 5 = totally satisfied.
[d] From 1 = very poor to 5 = very good.
[e] From 1 = never to 5 = very often.
[f] From 1 = never to 5 = always.
[g] From 1 = not at all satisfied to 5 = totally satisfied.
[h] From 1 = very unhappy to 5 = very happy.

Measures

Socio-demographic indicators

Respondents completed questions about their gender, age, household size and composition (single, with a spouse, with children, with a spouse and children, living in a nursing home), professional status (retired or professionally active), years of education, and monthly net income in euros per household (less than 1000, 1000–1600, 1601–2200, 2201–3000, 3001–5000, over 5000). The detailed characteristics of respondents are presented in Table 1.

Holidays and holiday-related activities

Respondents answered questions about the frequency and mean duration of holidays taken during the past 12 months, as well as their characteristics: destination, organized travel (yes or no), travel companion(s) (alone, partner, others), the frequency of participation in various activities (physical, social, cognitive, relaxing)[5] and the degree of perceived health benefits. *The frequency of holidays* in the past 12 months was measured on a five-point scale ranging from 1 (no holiday) to five (over 3 holidays). The *mean duration* of holidays was the ratio between the total number of nights spent on holidays and the number of holidays taken in the past 12 months. *The frequency of participation in various activities performed during the last holiday* was assessed on a five-point scale ranging from one (never) to five (very often). Finally, *the degree of perceived health benefits* related to the last holiday ('Do you think that your last holiday has had a positive impact on your health?') was assessed on a five-point scale ranging from one (not at all) to five (a lot). The detailed characteristics of holidays in senior tourists are presented in Table 2.

Wellbeing

As in a previous study (Galinha & Pais-Ribeiro, 2012), we created a global measure of wellbeing based on the definition suggested by Diener et al. (2003). We created a composite score of wellbeing including both cognitive (life satisfaction) and affective components (happiness). *Life satisfaction* was assessed by adopting the French version (Blais, Vallerand, Pelletier, & Brière, 1989) of the Satisfaction with Life Scale ([SWLS], Diener, Emmons, Larsen, & Griffin, 1985), a measure widely used in studies on the relationships between tourism experiences and wellbeing (e.g. Chen, Petrick, & Shahvali, 2016; Ferrer et al., 2016; Gilbert & Abdullah, 2004; McCabe & Johnson, 2013; Staats & Pierfelice, 2003; Sirgy et al., 2011). Respondents answered to five items on a five-point scale ranging from one (not at all agree) to five (strongly agree). The five items of this questionnaire yielded a Cronbach's alpha of .87 and they were averaged to obtain the SWLS score. *Happiness* was measured by the first item ('To what extent do you feel happy or unhappy?') adopted from the Happiness Measure ([HM], Fordyce, 1988). Respondent answered on a five-point scale ranging from one (very unhappy) to five (very happy). The HM index measures the affective component of wellbeing and is well known to have good psychometric properties in terms of reliability, construct and discriminative validity (Jarden, 2011). According to Diener (1984), HM should be more adopted in studies using wellbeing measures. SWLS and HM scales are reported in psychology literature as reliable assessments of wellbeing (Jarden, 2011). In this study, the cognitive (life satisfaction) and affective (happiness) components were highly inter-correlated, ($r = .69$, $p < .001$), reinforcing the utility of combining these dimensions into a unique composite score of wellbeing.

Table 2. Characteristics of holidays in senior tourists.

Variables	M	SD
Holidays in the last 12 months		
Frequency of holidays in the last 12 months		
1 holiday (%)	37.6	
2 holidays (%)	30.9	
3 holidays (%)	16.6	
Over 3 holidays (%)	14.9	
Mean duration of each holiday in the last 12 months (days)	6.85	5.77
Last holiday		
Destination		
France (%)	47	
Mediterranean Europe (Spain, Italy, Greece, etc.) (%)	23	
Belgium (%)	12	
Other countries in Europe (Netherlands, Germany, etc.) (%)	11	
Rest of the world (Africa, etc.) (%)	7	
Organized travel		
Yes (%)	19.9	
No (%)	80.1	
Travel Companion(s)		
Alone (%)	7.4	
Partner (%)	69	
Others (%)	23.6	
Perceived health benefits of the last holiday[a]	4.10	1.00
Physical activities[b]	3.73	1.23
Social activities[b]	2.28	1.33
Cognitive activities[b]	3.41	1.26
Relaxing activities[b]	2.94	1.21

Notes: $N = 2661$.
M: mean, SD: standard deviation.
[a]From 1 = not at all to 5 = a lot.
[b]From 1 = never to 5 = very often.

Self-rated health

Respondents provided multiple indicators of health: general health, the number of health symptoms, the degree of functional limitations, and self-rated memory problems. *General health* ('How do you rate your overall health?') was assessed on a five-point scale ranging from one (very poor) to five (very good). The number of *health symptoms* ('Which health problems do you have currently?') was based on a list of the 11 most common diseases in both genders of the Belgian population over 65 years of age (i.e. hypertension, cholesterol, diabetes, thyroid problems, arthritis, arthritis, back pain, urinary disorder, eye disorders, osteoporosis, prostate disorder) and other health problems added for the current study (i.e. asthma, insomnia, difficulty moving, other). The degree of *functional limitations* ('Have you been limited for at least six months because of a health problem in activities that you practice everyday?') was assessed on a five-point scale ranging from one (never) to five (very often). Finally, the degree of *self-rated memory failures* ('Do you have memory problems in your everyday life?') was assessed on a five-point scale ranging from one (never) to five (always). These indicators were adapted from the Belgian Health Survey (Scientific Institute of Public Health, 2013) except for the indicator of self-rated memory failures (Van der Linden, Wyns, von Frenckell, Coyette, & Seron, 1989).

General physical activity

Respondents reported the frequency of physical activity in their everyday life ('Do you practice physical activity such as sports, gardening, walking, etc.') on a six-point scale (one = less than four times a year, two = two to three times a year, three = one to two times a month, four = one to two times a week, five = three to six times a week, six = over six times a week). This item was adapted from the Dijon Physical Activity Score that has been validated in French in healthy seniors (Robert et al., 2004).

Social relations

Two indicators were used to assess the frequency and quality of social relations. *The frequency of social contacts* was assessed by one item ('Usually, how often do you have contact [face-to-face, telephone] with other people such as your spouse, children, friends, acquaintances, personal caregiver, etc.?') on a six-point scale (one = never, two = a few times a year, three = one to four times a month, four = one to three times a week, five = three to six times a week, six = each day of the week). *Quality of social contacts* was assessed by one item related to perceived social support ('How satisfied are you with the support of these people in your everyday life?') on a five-point scale ranging from one (not at all satisfied) to five (totally satisfied). These questions were adapted from the Belgian Health Survey (Scientific Institute of Public Health, 2013).

Data analysis

The data analyses were performed using IBM 23.3 SPSS Statistics (IBM Corp. 2015). The first research question was examined by conducting independent-samples t-tests (Chi-Squared tests for categorical variables) to compare senior tourists who went at least once on holidays in the last 12 months and senior non-tourists on socio-demographic indicators, self-rated health, general physical activity, and social relations. For the second research question, a one-way analysis of covariance (ANCOVA) was performed to compare scores of wellbeing in senior tourists and non-tourists, while taking into account the control variables (socio-demographic indicators, self-rated health, general physical activity, and social relations). Finally, the third research question was explored using a hierarchical multiple regression model with the following blocks of predictors: (1) control variables (age, gender, education years, income, general health, health symptoms, functional limitations and self-rated memory failures, general physical activity and frequency and quality of social relations), and (2) holiday-related factors (frequency, mean duration, frequency of physical, social, cognitive and relaxing activities and degree of perceived health benefits). All collinearity statistics were within accepted limits (Field, 2009; Myers, 1990), with the tolerance values for each predictor between .60 and .97 and the VIF values between 1.04 and 1.67. Finally, the assumption of the independence of errors was met with a value of 1.91 for the Durbin-Watson statistic, which is within the acceptable range (between 1.50 and 2.50, as suggested by Hair, Anderson, Tatham, & Black, 1998).

Results

How do the profiles of senior tourists and senior non-tourists differ?

In the present sample, 2661 respondents (64.4% tourists) had taken at least one holiday during the last 12 months before the survey, and 1469 (35.6%) were considered non-tourists. The detailed characteristics of holidays in senior tourists are presented in Table 2.

The results of comparisons between senior tourists and senior non-tourists (Table 3) revealed significant differences in terms of socio-demographic indicators (except for the gender), general physical activity, and social relations. Senior tourists were significantly younger, better educated, wealthier, more engaged in physical and social activities in their everyday life, and more satisfied with their social relations than senior non-tourists. In addition, senior tourists reported significantly higher scores on general health and lower scores on health symptoms, functional limitations, and self-rated memory failures.

Do senior tourists have higher scores of wellbeing than senior non-tourists, after adjusting for control variables?

There was a significant difference between senior tourists and senior non-tourists in terms of wellbeing, after adjusting for control variables (socio-demographic indicators, self-rated health, general physical activity and social relations), $F(1, 4015) = 123,94$, $p < .001$, $\eta p^2 = .03$. The senior tourists had higher scores of wellbeing ($M = 3.78$, $SD = .61$) than senior non-tourists ($M = 3.31$, $SD = .77$) over and above the control variables.

What are the holiday-related factors that significantly account for wellbeing in senior tourists?

As showed in Table 4, the results of the hierarchical multiple regression showed that holiday measures explained an additional 4% of the variation of wellbeing over and above the other significant predictors in the first block, $[F(7, 2569) = 21.47, p < .001]$. Block two showed that the more often seniors went on holidays, the more they perceived that the holidays had health benefits. In addition, the more they practiced holiday-related social and cognitive activities, the higher their wellbeing scores were. However, the duration of the holidays and the frequency of holiday-related physical and relaxing activities were not significantly associated with wellbeing of senior tourists.

Table 3. Comparisons between senior non-tourists and senior tourists.

Variables	Senior non-tourists			Senior tourists			df	t/χ^2	p	d
	n	M/%	SD	n	M/%	SD				
Socio-demographic indicators										
Gender (% of men)	916	62.4		1584	59.5		1	$\chi^2 = 3.17$	.08	
Age	1469	69.06	6.28	2661	67.74	5.46	2689.74	6.77	<.001	.22
Education years	1469	11.62	4.26	2661	13.15	4.12	2941.49	−11.21	<.001	.37
Net monthly income per household	1449	2.99	1.07	2602	3.70	1.14	3156.61	−20.01	<.001	.64
General physical activity	1469	3.75	1.56	2661	4.27	1.17	2393	−11.14	<.001	.38
Social relations										
Frequency of social contact	1451	6.19	3.76	2640	7.53	4.63	3526.86	−10.01	<.001	.32
Quality of social contact	1469	3.85	.98	2661	4.19	.85	2670.49	−11.07	<.001	.37
Self-rated health										
General health	1469	3.42	.87	2661	3.81	.76	2686.98	−14.60	<.001	.48
Health symptoms	1469	2.42	1.88	2661	2.08	1.66	2725.93	5.71	<.001	.19
Functional limitations	1460	2.85	1.33	2647	2.30	1.18	2711.10	13.21	<.001	.44
Self-rated memory failures	1469	2.58	.92	2661	2.45	.85	2843.92	4.45	<.001	.15

Notes: *n*: sample size, *M*: mean, *SD*: standard deviation, *df*: degrees of freedom, *t*: paired-sample *t*-tests, χ^2: Chi-squared tests, *p*: p-value, *d*: Cohen's d (effect size).

Table 4. Hierarchical multiple regression analysis predicting wellbeing in senior tourists.

	Wellbeing					
	Model 1			Model 2		
Steps and predictors variables	B	SE B	β	B	SE B	β
Step 1: Control variables						
Age	.01	.00	.10***	.01	.00	.11***
Gender	−.06	.02	−.05**	−.09	.02	−.07***
Years of education	.00	.00	.01	−.00	.00	−.01
Income	.10	.01	.19***	.09	.01	.16***
General health	.25	.02	.31***	.22	.02	.27***
Health symptoms	−.02	.01	−.06**	−.02	.01	−.06***
Functional limitations	−.01	.01	−.02	−.01	.01	−.03
Self-rated memory failures	−.09	.01	−.13***	−.08	.01	−.12***
General physical activity	.04	.01	.07***	.03	.01	.05**
Frequency of social contacts	.05	.01	.07***	.05	.01	.06***
Quality of social contacts	.14	.01	.18***	.13	.01	.17***
Step 2: Holiday-related factors						
Frequency				.07	.01	.13***
Mean duration				.00	.00	.00
Physical activities				−.01	.01	−.02
Social activities				.03	.01	.06***
Cognitive activities				.02	.01	.04**
Relaxing activities				.01	.01	.01
Perception of health benefits				.07	.01	.12***
R^2		.31***			.35***	
ΔR^2					.04***	

Notes: Gender: 1 = men, 2 = women.
SE: standard error, B: unstandardized regression coefficient, β: standardized regression coefficient, R^2: R-squared, ΔR^2: R-squared change.
*$p \leq .05$; **$p \leq .01$; ***$p \leq .001$.
Valid N = 2588 (listwise).

Additional predictors of wellbeing

As presented in Table 4, the final model of hierarchical regression analyses accounted for 35% of the total variance of wellbeing. Being older, being a man, having a higher income, feeling healthy, feeling physically active, having more social contacts, and being satisfied with the support of one's social contacts were significantly related to higher wellbeing.

Discussion

As expected, the results showed that when compared to senior non-tourists, senior tourists were significantly younger, better educated, wealthier, more engaged in physical and social activities in everyday life and reported better health. These findings confirm those presented in recent studies that highlighted significant differences between older tourists and non-tourists (Ferrer et al., 2016; Gu et al., 2016). Second, the present study found that senior tourists had higher levels of wellbeing than senior non-tourists, after adjusting for control variables (socio-demographic indicators, self-rated health, general physical activity and social relations). Third, we found that holiday-related factors predicted greater wellbeing in senior tourists over and above the role of relevant covariates. Our results are in line with previous studies that showed a positive association between holiday participation and health and wellbeing in seniors (e.g. Ahn & Janke, 2011; Ferrer et al., 2016; Gu et al., 2016; Kim et al., 2015; Nimrod & Rotem, 2012; Wei & Milman, 2002). This study provided a better understanding on the relationships between holidays and wellbeing by exploring the unique contribution of various holiday-related factors on wellbeing. Among the holiday-related factors, higher holiday frequency, frequency of social and cognitive activities, and degree of perceived health benefits were associated with higher levels of wellbeing.

The results showed differences in the contribution of the multiple holiday-related predictors of wellbeing. The frequency of holidays was found to be a significant predictor of wellbeing while mean duration was not. In other words, our findings suggest that it could be more beneficial to go on holidays several times over a year rather than to go once for a longer duration. Although multiple studies showed that the frequency of leisure activities was positively associated with seniors' wellbeing (e.g. Kelly et al., 2014; Kuykendall, Tay, & Ng, 2015; Windle, Hughes, Linck, Russell, & Woods, 2010), to the best of our knowledge, our study is the first to highlight that the frequency of holidays is associated with seniors' wellbeing. A possible interpretation is that each holiday is associated with developing a specific personal project, which generates positive affect before departure (e.g. Chen & Petrick, 2013; Hagger & Murray, 2014; Moal–Ulvoas & Taylor, 2014; Lawton, Moss, Winter, & Hoffman, 2002; Nawijn, Marchand, Veenhoven, & Vingerhoets, 2010). In this context, the more often seniors go on holidays, the more they develop personal projects and associated positive emotions in their everyday life.

Among the different types of activities practiced on holidays, social activities (e.g. group activities) represented the most important predictor of respondents' wellbeing. This finding supports other studies that have shown the benefits of holiday-related social activities (e.g. new social interactions, social sharing, satisfaction with social contacts, restoring and/or deepening relationships with others) on the reduction of loneliness (Toepoel, 2013), social inclusion (Ferrer et al., 2016; Gu et al., 2016; McCabe & Johnson, 2013; Minnaert et al., 2009; Morgan et al., 2015) and senior tourists' wellbeing (Ferrer et al., 2016; Hunter-Jones & Blackburn, 2007; Moal–Ulvoas & Taylor, 2014; Nimrod & Rotem, 2012). Many studies have demonstrated that the participation in social activities improves seniors' wellbeing (e.g. Adams, Leibbrandt, & Moon, 2011; Dupuis, 2008; Levasseur et al., 2015; Litwin & Shiovitz-Ezra, 2011; McAuley et al., 2000; Menec, 2003; Tsai & Wu, 2005). Indeed, seniors are more likely to experience greater loneliness in their everyday life compared to younger individuals, especially due to retirement and/or other factors associated with the ageing process such as the death of close ones (Gibson & Singleton, 2012). Social participation represents an adaptive strategy to counter social deficits related to ageing (e.g. Balderas-Cejudo et al., 2017; Silverstein & Parker, 2002).

Cognitive activities performed on holidays (e.g. reading, orienting, planning) represented another significant predictor of senior tourists' wellbeing. To the best of our knowledge, this is also the first study showing a positive association between cognitive activities practiced during the holidays and wellbeing of senior tourists. This result is consistent with previous studies observing this association in everyday life (e.g. Allward, Dunn, Forshaw, Rewston, & Wass, 2017; Chang et al., 2014; Olazaran et al., 2010; Paillard-Borg, Wang, & Winblad, 2009). For instance, Paillard-Borg et al. (2009) showed that cognitive activities (e.g. writing, playing music) enhanced the wellbeing of seniors more than did physical and recreational activities (e.g. watching TV). However, the underlying mechanisms of the effect of cognitive activities on wellbeing are not well known. As indicated by Gu et al. (2016), we can suggest that multiple cognitive activities practiced during the holidays (e.g. orienting oneself in a new town, activity planning, learning new information) may cognitively stimulate senior tourists, which generates a greater sense of wellbeing. This idea should be further explored in future research.

Finally, our results also revealed that the perception of health benefits related to holidays significantly predicted senior tourists' wellbeing. Chen and Petrick (2014) showed that the more tourists perceive that holidays have health benefits, the more they go on holidays. These authors suggest that the perception of health benefits associated with holiday experiences increases the perceived importance of holidays in everyday life (i.e. increased attention to information and discussions about future holidays), thereby improving the likelihood of going on holidays. As described above, several studies have shown that the frequency of leisure activities is associated with seniors' wellbeing. Therefore, it may be the case that in our study, the more senior tourists perceived the health benefits of various leisure activities practiced on holidays, the more they engaged in these activities, and the happier they were as a result.

The findings have allowed to better grasp the factors accounting for wellbeing during ageing from a psychological perspective and their link with holidays. The results support the claims of social tourism researchers and practitioners to promote social tourism and to facilitate holiday participation, notably through social tourism provision. The comparison of seniors going on holidays and non-tourists clearly points out the benefits of holidays. This suggests that seniors should be made aware of the protective and beneficiary aspects for their health and wellbeing. Additionally, based on our findings, the tourism industry should develop more services adapted to seniors in order to comply with their specific needs and preferences. To this end, guidelines about the specific needs of seniors and their preferences in terms of types of activities could be created for tourism providers. At the societal level, the findings of this study could serve as evidence for encouraging public authorities to facilitate the access of holiday resources for low-income seniors. As showed by some studies, this strategy may have a beneficial effect on various aspects of the health and wellbeing of those people. Through tourism, seniors get the opportunity for social encounters and meet new people. Moreover, holidays help them to get out of their isolation, and be more physically active. They can forget about financial problems and others sources of worry during the time of the trip and focus their attention on more positive sides of their life (Ferrer et al., 2016; Medaric, Gabruc, & Sedmak, 2016; Morgan et al., 2015). Recent literature on social tourism widely recognizes potential benefits of holidays as an integral part of active ageing and a major form of intervention in social care (Medaric et al., 2016; Minnaert & Schapmans, 2009).

The results of the current study should be considered in light of several limitations. Firstly, our sample is not representative of the Belgian population of elderly people, especially considering the method of recruitment performed exclusively by a single health insurance company (albeit the largest in Belgium) and the online nature of the questionnaires. Indeed, 37% of Belgian adults aged between 65 and 75 years have never used Internet (Statbel, 2017b), which excluded a large proportion of this population from our study. Moreover, the sample was dominated by men whereas the opposite is found in Belgian population aged over 65 years (Statbel, 2017a). Secondly, the cross-sectional design does not allow us to draw any conclusions regarding the directionality of the association between holiday participation and wellbeing. Thirdly, that is possible that the validity of the measures has been affected by the retrospective nature of this study that could have generated memory bias. Finally, our findings should not be considered as an agenda for policy. For example, one should not only focus on the practice of social and cognitive activities during holidays simply because these were the activities with the strongest associations with wellbeing. In addition, the ageing population is known to be a heterogeneous group (Lowsky, Olshansky, Bhattacharya, & Goldman, 2014), especially in terms of personal needs and interests. As such, seniors need to find personally fulfilling leisure experiences that give meaning to their lives (Gibson & Singleton, 2012).

Future research should consider longitudinal designs to explore the influence of a holiday on seniors' health and wellbeing. Additionally, pre–post designs could eliminate potential baseline differences and increase the validity of measures by limiting memory bias. There is also a need to investigate the mechanisms that underlie the relationship between the different types of social and cognitive activities practiced on holidays and senior tourists' wellbeing. Finally, future research could investigate how long the benefits of a holiday last for in seniors.

Conclusion

The research aims have been met by providing valuable data on holiday-related predictors of seniors' wellbeing. The results of this study showed, as expected, that senior tourists had a different profile than senior non-tourists: they were younger, more educated, wealthier, and healthier. In addition, their levels of wellbeing were higher compared to senior non-tourists, after controlling for socio-demographic indicators, health, physical activity, and social relations. The findings also showed that holidays significantly contribute to higher wellbeing scores over and above the role of relevant covariates. The analyses of significant holiday-related predictors of wellbeing showed that the more often seniors went on holidays, the more they practiced holiday-related social and cognitive activities and the more they perceived that holidays had health benefits, the higher their wellbeing scores.

Notes

1. Holidays refer to all journeys for pleasure including at least four consecutive nights away from home (World Tourism Organization [WTO], 1995).
2. Bien-être, Emploi, Santé et Tourisme social in collaboration with Université libre de Bruxelles (ULB) and funded by the Walloon region (Belgium) – Germaine Tillon funding on social innovation.
3. The Katz Index of Independence in Activities of Daily Living assesses the ability to perform activities of daily living independently. Patients are scored on a four-point scale for independence in each of six functions (bathing, dressing, toileting, transferring, continence, feeding). The scores range from one (totally independent) to four (totally dependent). People with a score of three or four on at least two criteria of the scale were excluded from the mailing list.
4. This percentage corresponds to the average level of responses obtained by the Mutualité Chrétienne in previous surveys.
5. The different activities were presented using specific examples: physical activities (e.g. visits, walks, gymnastics, aqua gym, dance), social activities (e.g. workshops, group games, group excursions), cognitive activities (e.g. reading, word games, itinerary planning) and relaxing activities (e.g. getting a massage, lying on a deckchair, taking a nap).

Acknowledgments

This work was supported by the Germaine Tillion Grant [number 1318182] for Social Innovation (BEST project) in collaboration with Université libre de Bruxelles (ULB) from the Walloon region (Belgium) and the Belgian Fund for Scientific Research (F.R.S.-FNRS) accorded to Olivier Luminet. We thank Mutualité Chrétienne for the recruitment of participants. We would also like to thank Betty Chang for her proofreading of the paper, and Aurélie Van der Haegen, Djouaria Ghilani, Elke Vlemincx, Giorgia Zamariola, Jessica Morton and Valérie Broes for their helpful comments on the manuscript.

Disclosure statement

No potential conflict of interest was reported by the authors.

Funding

This work was supported by the Germaine Tillion (walloon region) [grant number 1318182] for Social Innovation (BEST project) in collaboration with Université libre de Bruxelles (ULB) from the Walloon region (Belgium) and the Belgian Fund for Scientific Research (F.R.S.-FNRS) accorded to Olivier Luminet.

References

Adams, K. B., Leibbrandt, S., & Moon, H. (2011). A critical review of the literature on social and leisure activity and wellbeing in later life. *Ageing & Society, 31*, 683–712. doi:10.1017/S0144686X10001091

Ahn, Y.-J., & Janke, M. C. (2011). Motivations and benefits of the travel experiences of older adults. *Educational Gerontology, 37*, 653–673. doi:10.1080/03601271003716010

Alén, E., Losada, N., & de Carlos, P. (2017). Profiling the segments of senior tourists throughout motivation and travel characteristics. *Current Issues in Tourism, 20*(14), 1454–1469. doi:10.1080/13683500.2015.1007927

Allward, C., Dunn, R., Forshaw, G., Rewston, C., & Wass, N. (2017). Mental wellbeing in people with dementia following cognitive stimulation therapy: Innovative practice. *Dementia (basel, Switzerland)*, 1–9. doi:10.1177/1471301217722443

Balderas-Cejudo, M. A., Leeson, G. W., & Urdaneta, E. (2017). Social tourism: Towards and active healthy ageing. *Open Access Journal of Gerontology & Geriatric Medicine, 1*(3), 555563. doi:10.19080/OAJGGM.2017.01.555563

Blais, M. R., Vallerand, R. J., Pelletier, L. G., & Brière, N. M. (1989). L'échelle de satisfaction de vie: Validation canadienne-française du "satisfaction with life scale" [The satisfaction with life scale: French-Canadian validation of 'satisfaction with life scale']. *Canadian Journal of Behavioural Science / Revue Canadienne des Sciences du Comportement, 21*(2), 210–223.

Caradec, V., & Petite, S. (2008). Voyages organisés à la retraite et lien social [Travel organized at retirement and social link]. *Retraite et Société, 4*(56), 139–168.

Chang, P. J., Wray, L., & Lin, Y. (2014). Social relationships, leisure activity, and health in older adults. *Health Psychology, 33*(6), 516–523. doi:10.1037/hea0000049

Chen, C. C., & Petrick, J. F. (2013). Health and wellness benefits of travel experiences: A literature review. *Journal of Travel Research, 52*(6), 709–719. doi:10.1177/0047287513496477

Chen, C. C., & Petrick, J. F. (2014). The roles of perceived travel benefits, importance, and constraints in predicting travel behavior. *Journal of Travel Research, 55*(4), 1–14. doi:10.1177/0047287514563986

Chen, C. C., Petrick, J. F., & Shahvali, M. (2016). Tourism experiences as a stress reliever: Examining the effects of tourism recovery experiences on life satisfaction. *Journal of Travel Research, 55*(2), 150–160. doi:10.1177/0047287514546223

Chen, L. J., Stevinson, C., Ku, P. W., Chang, Y. K., & Chu, D. C. (2012). Relationships of leisure-time and non-leisure- time physical activity with depressive symptoms: A population-based study of Taiwanese older adults. *International Journal of Behavioral Nutrition and Physical Activity*, 9 (28). doi:10.1186/1479-5868-9-28

Dann, G. M. S. (2001). Senior tourism and quality of life. *Journal of Hospitality and Leisure Marketing*, 9(1), 5–15.

De Bloom, J., Geurts, S. A. E., Taris, T. W., Sonnentag, S., de Weerth, C., & & Kompier, M. A. J. (2010). Effects of vacation from work on health and well-being: Lots of fun, quickly gone. *Work & Stress*, 24(2), 196–216. doi:10.1080/02678373.2010.493385

Diekmann, A., & McCabe, S. (2011). Systems of social tourism in the European Union: A critical review. *Current Issues in Tourism*, 14(5), 417–430. doi:10.1080/13683500.2011.568052

Diekmann, A., & McCabe, S. (2016). Social tourism and health. In M. K. Smith, & L. Puczko (Eds.), *The Routledge handbook of health tourism* (pp. 103–112). London and New-York: Routledge.

Diener, E. (1984). Subjective well-being. *Psychological Bulletin*, 95(3), 542–575. doi:10.1037/0033-2909.95. 3.542

Diener, E., Emmons, R. A., Larsen, R. J., & Griffin, S. (1985). The satisfaction with life scale. *Journal of Personality Assessment*, 49, 71–75. doi:10.1207/s15327752jpa4901_13

Diener, E., Oishi, S., & Lucas, R. E. (2003). Personality, culture, and subjective well-being: Emotional and cognitive evaluations of life. *Annual Review of Psychology*, 54, 403–425. doi:10. 1146/annurev.psych.54.101601.145056

Dolnicar, S., Laesser, C., & Matus, K. (2009). Online versus paper - format effects in tourism surveys. *Journal of Travel Research*, 47(3), 295–316. doi:10.1177/0047287508326506

Dolnicar, S., Yanamandram, V., & Cliff, K. (2012). The contribution of vacations to quality of life. *Annals of Tourism Research*, 39(1), 59–83. doi:10.1016/j.annals.2011.04.015

Dupuis, S. L. (2008). Leisure and ageing well. *World Leisure Journal*, 50(2), 91–107. doi:10.1080/04419057.2008.9674538

Ferrer, J. G., Sanz, M. F., Ferrandis, E. D., McCabe, S., & Garcia, J. S. (2016). Social tourism and healthy ageing. *International Journal of Tourism Research*, 18, 297–307. doi:10.1002/jtr.2048

Field, A. P. (2009. *Discovering statistics using SPSS: And sex and drugs and Rock 'n' Roll* (3th ed.). London: Sage publications.

Fordyce, M. W. (1988). A review of research on the happiness measures: A sixty second index of happiness and mental health. *Social Indicators Research*, 20, 355–381. doi:10.1007/BF00302333

Galinha, I., & Pais-Ribeiro, J. L. (2012). Cognitive, affective and contextual predictors of subjective wellbeing. *International Journal of Wellbeing*, 2(1), 34–53. doi:10.5502/ijw.v2i1.3

Gibson, H. J., & Singleton, J. F. (2012). *Leisure and aging: Theory and practice*. Champaign, IL: Human Kinetics.

Gilbert, D., & Abdullah, J. (2004). Holidaytaking and the sense of well-being. *Annals of Tourism Research*, 31(1), 103–121. doi:10.1016/j.annals.2003.06.001

Gu, D., Zhu, H., Brown, T., Hoenig, H., & Zeng, Y. (2016). Tourism experience and self-rated health among older adults in China. *Journal of Aging and Health*, 28(4), 675–703. doi:10.1177/0898264315609906

Hagger, C., & Murray, D. (2014). Anticipating a flourishing future with tourism experiences. In S. Filep, & P. Pearce. (Eds.), *Tourist experience and fulfilmen: Insights from positive psychology* (pp. 238–261). Page: Routledge.

Hair, J. F., Anderson, R. E., Tatham, R. L., & Black, W. C. (1998). *Multivariate data analysis*. Prentice-Hall: Pearson.

Havighurst, R. J. (1961). Successful Aging1. *The Gerontologist*, 1(1), 8–13. doi:10.1093/geront/1.1.8

Higgins-Desbiolles, F. (2006). More than an industry: Tourism as a social force. *Tourism Management*, 26(6), 1192–1208. doi:10.1016/j.tourman.2005.05.020

Hunter-Jones, P., & Blackburn, A. (2007). Understanding the relationship between holiday-taking and self-assessed health: An exploratory study of senior tourism. *International Journal of Consumer Studies*, 31, 509–516. doi:10.1111/j.1470-6431.2007.00607.x

Jarden, A. (2011). *Positive psychological assessment: a practical introduction to empirically validated research tools for measuring wellbeing* [PDF file]. Retrieved from http://www.positivepsychology.

org.nz/uploads/3/8/0/4/3804146/workshop_4_-_dr_aaron_jarden_-_positive_psychological_ assessment_workbook.pdf

Jia, B. B., Yang, Z. Y., Gen, X., Lyu, Y. D., Wen, X. L., Xu, Y. H., … Wang, G. F. (2016). Health effect of forest bathing trip on elderly patients with chronic obstructive pulmonary disease. *Biomedical Environmental Science*, *29*(3), 212–218. doi:10.3967/bes2016.02

Katz, S., Down, T. D., Cash, H. R., & Grotz, R. C. (1970). Progress in the development of the index of ADL. *The Gerontologist*, *10*(1), 20–30. doi:10.1093/geront/10.1_Part_1.20

Kelly, M. E., Loughrey, D., Lawlor, B. A., Robertson, I. H., Walsh, C., & Brennan, S. (2014). The impact of cognitive training and mental stimulation on cognitive and everyday functioning of healthy older adults: A systematic review and meta-analysis. *Aging Research Reviews*, *15*, 28– 43. doi:10.1016/j.arr.2014.05.002

Kim, H., Woo, E., & Uysal, M. (2015). Tourism experience and quality of life among elderly tourists. *Tourism Management*, *46*, 465–476. doi:10.1016/j.tourman.2014.08.002

Kuykendall, L., Tay, L., & Ng, V. (2015). Leisure engagement and subjective well-being: A quanti- tative review. *Psychological Bulletin*, *141*, 364–403. doi:10.1037/a0038508

Lawton, M. P., Moss, M. S., Winter, L., & Hoffman, C. (2002). Motivation in later life: Personal projects and well-being. *Psychology and Aging*, *17*(4), 539–547.

Levasseur, M., Généreux, M., Bruneau, J. F., Vanasse, A., Chabot, E., Beaulac, C., & Bédard, M. M. (2015). Importance of proximity to resources, social support, transportation and neighborhood security for mobility and social participation in older adults: Results from a scoping study. *BMC Public Health*, *15*(503). doi:10.1186/s12889-015-1824-0

Litwin, H., & Shiovitz-Ezra, S. (2011). Social network type and subjective well-being in a national sample of older Americans. *The Gerontologist*, *51*(3), 379–388. doi:10.1093/geront/gnq094

Lowsky, D. J., Olshansky, S. J., Bhattacharya, J., & Goldman, D. P. (2014). Heterogeneity in healthy aging. *Journals of Gerontology Series A: Biological Sciences and Medical Sciences*, *69*(6), 640–649. doi:10.1093/gerona/glt162

McAuley, E., Blissmer, B., Marquez, D., Jerome, G. J., Kramer, A. F., & Katula, J. (2000). Social relations, physical activity and well-being in older adults. *Preventive Medicine*, *31*, 608–617. doi:10.006/pmed.2000.0740

McCabe, S., & Johnson, S. (2013). The happiness factor in tourism: Subjective well-being and social tourism. *Annals of Tourism Research*, *41*(1), 42–65. doi:10.1016/j.annals.2012.12.001

Medaric, Z., Gabruc, J., & Sedmak, M. (2016). Social tourism benefits for seniors. *Academia Turistica*, *9*(2), 113–116.

Menec, V. (2003). The relationship between everyday activities and successful aging: A 6-year longi- tudinal study. *The Journals of Gerontology Series B: Psychological Sciences and Social Sciences*, *58* (2), S74–S82.

Minnaert, L., Maitland, R., & Miller, G. (2009). Tourism and social policy - the value of social tourism. *Annals of Tourism Research*, *36*(2), 316–334. doi:10.1016/j.annals.2009.01.002

Minnaert, L., & Schapmans, M. (2009). Tourism as a form of social intervention: The holiday par- ticipation centre in Flanders. *Journal of Social Intervention: Theory and Practice*, *18*(3), 42–61. doi:10.18352/jsi.171

Mitas, O., Yarnal, C., Adams, R., & Ram, N. (2012). Taking a 'peak' at leisure travelers' positive emotions. *Leisure Sciences*, *34*(2), 115–135. doi:10.1080/01490400.2012.652503

Moal–Ulvoas, G., & Taylor, V. A. (2014). The spiritual benefits of travel for senior tourists. *Journal of Consumer Behaviour*, *13*(6), 453–462. doi:10.1002/cb.1495

Morgan, N., Pritchard, A., & Sedgley, D. (2015). Social tourism and well-being in later life. *Annals of Tourism Research*, *52*, 1–15. doi:10.1016/j.annals.2015.02.015

Myers, R. H. (1990. *Classical and modern regression application* (2nd ed.). Boston: Duxbury Press.

Nawijn, J. (2011). Determinants of daily happiness on vacation. *Journal of Travel Research*, *50*(5), 559–566. doi:10.1177/0047287510379164

Nawijn, J., Marchand, M., Veenhoven, R., & Vingerhoets, A. (2010). Vacationers happier, but most not happier after a holiday. *Applied Research in Quality of Life*, *5*(1), 35–47. doi:10.1007/s11482- 009-9091-9

Nimrod, G. (2008). Retirement and tourism themes in retirees' narratives. *Annals of Tourism Research*, 35(4), 859–878. doi:10.1016/j.annals.2008.06.001

Nimrod, G., & Kleiber, D. (2007). Reconsidering change and continuity in later life: Toward an innovation theory of successful aging. *International Journal of Aging and Human Development*, 65(1), 1–22. doi:10.2190/Q4G5-7176-51Q2-3754

Nimrod, G., & Rotem, A. (2010). Between relaxation and excitement: Activities and benefits gained in retirees' tourism. *International Journal of Tourism Research*, 12(1), 65–78. doi:10.1002/jtr.739

Nimrod, G., & Rotem, A. (2012). An exploration of the innovation theory of successful ageing among older tourists. *Ageing & Society*, 32, 379–404. doi:10.1017/S0144686X1100033X

Olazaran, J., Reisberg, B., Clare, L., Cruz, I., Pena-Casanova, J., Del Ser, T., ... Muniz, R. (2010). Nonpharmacological therapies in Alzheimer's disease: A systematic review of efficacy. *Dementia and Geriatric Cognitive Disorders*, 30(2), 161–178. doi:10.1159/000316119

Paillard-Borg, S., Wang, H. X., & Winblad, B. (2009). Pattern of participation in leisure activities among older people in relation to their health conditions and contextual factors: A survey in a Swedish urban area. *Ageing & Society*, 29, 803–821. doi:10.1159/000235576

Richards, G. (1999). Vacations and the quality of life: Patterns and structures. *Journal of Business Research*, 44(3), 189–198. doi:10.1016/S0148-2963(97)00200-2

Robert, H., Casillas, J. M., Iskandar, M., D'Athis, P., Antoine, D., Taha, S., ... Van Hoecke, J. (2004). Le score d'activité physique de Dijon: reproductibilité et corrélations avec l'aptitude physique de sujets sains âgés [The Dijon Physical Activity Score: reproducibility and correlation with exercise testing in healthy elderly subjects]. *Annales de Réadaptation et de Médecine Physique*, 47, 546–554. doi:10.1016/j.annrmp.2004.03.005

Schafer, J. L. (1999). Multiple imputation: A primer. *Statistical Methods in Medical Research*, 8(1), 3–15. doi:10.1177/096228029900800102

Scientific Institute of Public Health. (2013). *Health Interview Survey* [PDF file]. Retrieved from https://his.wiv-isp.be/fr/SitePages/Questionnaires.aspx

Silverstein, M., & Parker, M. G. (2002). Leisure activities and quality of life among the oldest old in Sweden. *Research on Aging*, 24, 528–547. doi:10.1177/0164027502245003

Sirgy, M. J. (2010). Toward a quality-of-life theory of leisure travel satisfaction. *Journal of Travel Research*, 49(2), 246–260. doi:10.1177/0047287509337416

Sirgy, M. J., Kruger, S. P., Lee, D. J., & Yu, G. B. (2011). How does a travel trip affect tourists' life satisfaction? *Journal of Travel Research*, 50(3), 261–275. doi:10.1177/0047287509337416

Smith, M. K., & Diekmann, A. (2017). Tourism and wellbeing. *Annals of Tourism Research*, 66, 1–13. doi:10.1016/j.annals.2017.05.006

Staats, S., & Pierfelice, L. (2003). Travel: A long-range goal of retired women. *The Journal of Psychology*, 137, 483–494. doi:10.1080/00223980309600630

Statbel. (2017a). *Structure de la population* [Excel file]. Retrieved from https://statbel.fgov.be/fr/themes/population/structure-de-la-population#panel-13

Statbel. (2017b). *Utilisation des TIC auprès des individus* [Excel file] Retrieved from https://statbel.fgov.be/fr/themes/menages/utilisation-des-tic-aupres-des-menages#panel-12

Toepoel, V. (2013). Ageing, leisure, and social connectedness: How could leisure help reduce social isolation of older people? *Social Indicators Research*, 113(1), 355–372. doi:10.1007/s11205-012-0097-6

Tsai, C. Y., & Wu, M. T. (2005). Relationship between leisure participation and perceived wellness among older persons in Taiwan. *Journal of ICHPER*, 41(3), 44–50.

Urry, J. (1995). *Consuming places*. London: Routledge.

Van der Linden, M., Wyns, C., von Frenckell, R., Coyette, G., & Seron, X. (1989). *Le Q.A.M. Questionnaire d'Auto-évaluation de la Mémoire [Memory self-evaluation questionnaire]*. Bruxelles: Editest.

Wei, S., & Milman, A. (2002). The impact of participation in activities while on vacation on seniors' psychological well-being: A path model application. *Journal of Hospitality and Tourism Research*, 26(2), 175–185. doi:10.1177/1096348002026002006

Williamson, J. (2016). Awareness of physical activity health benefits can influence participation and dose. *Sports and Medicine and Rehabilitation Journal*, 1(1), 1–7.

Windle, G., Hughes, D., Linck, P., Russell, I., & Woods, B. (2010). Is exercise effective in promoting mental well-being in older age? A systematic review. *Aging & Mental Health, 14*(6), 652–669. doi:10.1080/13607861003713232

World Tourism Organization. (1995). *Concepts, definitions, classifications for tourism statistics: Technical manual.* Spain: Author.

Accessible tourism and its benefits for coping with stress

Andreia Filipa Antunes Moura ⓘ, Elisabeth Kastenholz and Anabela Maria
Sousa Pereira

ABSTRACT

The present study had two main objectives: (i) to understand the relationship between tourism and stress-coping for individuals with disabilities, and (ii) to develop an empirical basis for therapeutic purposes and for improving new tourism products and policies, in a biopsychosocial framework. An empirical study was conducted using participants with disabilities ($N = 306$) who were assessed with the Leisure Coping Scale adapted to the Accessible Tourism context. The positive influence of tourism on these individuals' biopsychosocial dimensions of stress-coping is identified and discussed. Results suggest that accessible tourism should be recognized as a new stress-coping resource for disabled people, supporting the rebalancing of their personal and social resources, positively contributing to their health and well-being. These findings provide further evidence for the development of new tourism products targeted to a population with special needs, and for accurate policies of alternative therapeutic interventions in the context of their rehabilitation.

RESUMEN

El estudio tiene como objetivos principales los siguientes puntos: (i) para comprender la relación entre turismo y estrés coping para personas con discapacidad y para desarrollar una base empírica para propuestas terapéuticas para el desarrollo de nuevos productos de turismo. Un estudio empírico fue conducido en pacientes con discapacidad ($N = 306$) que han tenido acceso a la Escala Leisure Coping adaptada en un contexto de turismo accesible. La influencia positiva del turismo en estos individuos a nivel psicosocial fue identificada y discutida. Los resultados sugieren que el turismo accesible puede ser una nueva herramienta de estrés-coping para personas discapacitadas soportando el reequilibrio de sus vidas personales y sociales, contribuyendo para la salud y bienestar. Estos resultados pueden garantizar descubiertas para el desarrollo de nuevos productos para poblaciones específicas con necesidades especiales y aún

para nuevas y alternativas políticas terapéuticas en contextos de
rehabilitación.

RÉSUMÉ
Cette étude vise deux objectifs principaux: (i) comprendre la relation
entre le tourisme et l'adaptation au stress pour les personnes
handicapées, et (ii) développer une base empirique à des fins
thérapeutiques et à l'amélioration de nouveaux produits
touristiques dans un cadre biopsychosocial. Une étude empirique
a été menée auprès des participants ayant un handicap ($N = 306$)
qui ont été évalués à l'aide de l'échelle de loisir adaptée au
contexte du tourisme accessible. L'influence positive du tourisme
sur les dimensions biopsychosociales de l'adaptation au stress de
ces personnes est identifiée et discutée. Les résultats suggèrent
que le tourisme accessible devrait être reconnu comme une
nouvelle ressource pour les personnes handicapées qui résistent
au stress, en soutenant le rééquilibrage de leurs ressources
personnelles et sociales, contribuant positivement à leur santé et
à leur bien-être. Ces résultats fournissent des preuves pour le
développement de nouveaux produits touristiques focalisé sur
une population ayant des besoins spéciaux et pour une politique
précise des interventions thérapeutiques alternatives dans le
contexte de leur réadaptation.

摘要
这个研究有两个主要的目的: (1) 了解旅游和残疾人士压力处理
之间的关系, (2) 基于治疗目的以及改进新旅游产品和政策, 用
一个生物心理社会框架去建立一个实证的基础。此实证研究的对
象是 306 名残疾人士, 研究使用经调整适用于可进入旅游情景的
休闲处理测量 Leisure Coping Scale (LCS) 来对这些调研对象进行
评估。研究识别出以及讨论了旅游对这些人士处理压力的生物心
理社会方面的正面影响。   研究结果显示, 可进入旅游应该被视
为残疾人士的一个新的压力处理方法和资源, 支撑着他们的个人
和社会资源间的重新平衡, 正面地促进了他们的身心健康和幸
福。研究的这些发现为针对有特殊需求的人士的新旅游产品的发
展提供了进一步的实证证据, 也为在这些人士的复原时期的另类
的治疗型方法的正确政策提出进一步的数据证明。

Introduction

The interaction between stress, coping and leisure or recreation was consistently
studied, theoretically and empirically, by the prolific research of Lazarus and
Folkman (1984). However, only a few authors have contributed to the increase of
knowledge in the leisure field (Schneider & Iwasaki, 2003) and none, in particular,
in the field of tourism.

The literature proved the importance of leisure for health benefits and overall wellbeing and particularly for the recovery and rehabilitation of individuals in extreme
cases of loss or negative life events, such as the emergence of a disability (Kleiber, Reel,
& Hutchinson, 2008). Moreover, tourism is a recognized Human Right and an important
social activity (McCabe & Diekmann, 2015), increasingly related to what people do in their
free time and within recreational activities (Mannell & Kleiber, 1997). It thus seems that
social tourism is the appropriate framework for this research, as it refers to initiatives that

include individuals who would otherwise be excluded in tourism activities (Assipova & Minnaert, 2014). These include people living on a low income (McCabe, Minnaert, & Diekmann, 2012) or those socially disadvantaged, as is frequently the case of the disabled or the elderly (Walton, 2013).

So, in this study, we will analyse tourism as an activity developed in leisure time, whose intrinsic features and specific benefits are very important in coping, arousing feelings of freedom, autonomy and personal development, particularly relevant in minimizing and controlling stress (Folkman & Moskowitz, 2004). Our aim and research innovation objective is to fill this gap in knowledge, perceiving tourism as a coping resource for people with disabilities, one of the target groups of social tourism most frequently mentioned in the literature (Carneiro, Eusébio, Kastenholz, & Alvelos, 2013).

Regarding a more frequently studied type of disability – mobility impairment – and the fact that there is less-empirical evidence on other kinds of disability, the present paper also aims to eliminate part of this limitation, based on the empirical object established: people with physical and sensory disability, i.e. with mobility, visual and hearing impairments. The sampling limits also exclude the population with intellectual or mental disability, considering the complexity of the subject and its evaluation tools (Bramston & Fogarty, 1995), as it would not ensure reliable data and information collection (if they were themselves to answer the questions) or anything comparable to that (if it were their caregivers to respond).

Empirical evidence shows that leisure arouses positive feelings of freedom and personal development, which illustrates its potential to reduce and control stress by promoting social interaction, positive emotions and coping in the context of stressful situations (Folkman & Moskowitz, 2004). Since tourism is recognized as a source of positive impacts or benefits for an individual's overall well-being and long-term quality of life (La Placa & Corlyon, 2014; Neal, Uysal, & Sirgy, 2007), it can also be associated with stress reduction, which is why we decided to study the intrinsic aspects of this activity to minimize stress, from the perspective of people with disabilities.

This potential involvement of people with physical and sensory disability in tourism as a coping resource, is to be analysed, allowing us to establish two major research objectives. The first one, yielding a contribution to theory in the fields of tourism and psychology, namely in the domain of stress-coping, is to better understand the dynamic relationship between tourism and stress-coping amongst individuals with a disability. The second, more focused on practical implications, aims at an empirical basis for the development of relevant guidelines, not only for therapeutic purposes, but also for developing new, more inclusive tourism policies and products, rooted in a biopsychosocial approach. As suggested by the WTO Global Code of Ethics for World Tourism, 'Social Tourism and "Social Equity" is not a mere will of the wisp of idealists and marginalized pressure groups, but an aspiration shared at the highest levels of industry and governments' (Ryan, 2002, p. 19).

We will proceed by first presenting the literature review leading to the conceptual model proposed here. The methodology of data collection and analysis to then validate the model's hypotheses is presented as well as results of the empirical study. The final section is devoted to conclusions and a debate on the implications of results, some limitations of our approach as well as future avenues of research.

Literature review and conceptual model

Stress negatively affects peoples' lives in modern society, causing sickness and disease, diminishing daily performance and decreasing quality of life. The population with disabilities is probably more exposed to disadvantaged situations (Walton, 2013) or to different and more intense stress-inducing circumstances, which may cause instability – even more severe when affecting the individual's mental, psychological, emotional and even physical resources.

Accessible leisure tourism, adopted as the main focus of this research, induces positive effects on the development of its participants at various levels, which are even more intense for people with disabilities (Shaw & Coles, 2004). It may enhance positive emotions, thus serving the functions identified above, and may correspondingly be considered a stress-coping tool or a stress-control method, with particular relevance to the disabled population.

Research gaps in this area are evident, explaining the relevance of this study for the pursuit of knowledge and development of strategies to deal with stress for individuals with disabilities, with the potential of thereby contributing to health and quality of life of this population group (Mactavish & Iwasaki, 2005). Additionally, the involvement of people with disabilities in accessible tourism should enhance their social inclusion and permit the enjoyment of benefits that greatly affect their physical, social and mental well-being, thus positively contributing to their overall rehabilitation (Kastenholz, Eusébio, Moura, & Figueiredo, 2010).

Consequently, the relevance of this investigation is clear from (a) a scientific point of view, since it corresponds to a new area of knowledge in psychology, leisure and tourism, where research has given little attention to the role of leisure or recreational activities as stress-coping mechanism, particularly relevant for the disabled population (Iwasaki & Mannell, 2000; Kleiber, Hutchinson, & Williams, 2002; Lazarus, 1993; Zuzanek, Robinson, & Iwasaki, 1998), (b) a social perspective, as it promotes the inclusion of a population frequently affected by discrimination and prejudice, (c) a therapeutic angle, since it reveals means of enhancing the overall well-being of individuals with disability and (d) an economic standpoint, since it seeks to add arguments in favour of a new management paradigm in the tourism industry, namely 'accessible tourism' or 'tourism for all', stimulating loyal tourism demand.

Constructs established and selected were based on the studies of Lazarus and Folkman (1984), Iwasaki and Mannell (2000) and Mactavish and Iwasaki (2005), considering the population with disabilities as the research population (see Figure 1).

Through the presented conceptual model, it is possible to link the main topics under study, showing how it is suggested they interrelate and affect each other. What are the effects of tourism on people with disabilities? What are the benefits that tourism can provide to individuals with disabilities, so that they can cope with stress successfully? Do the individuals' type of disability and the socio-demographic profile influence the way they deal with stress through tourism? These questions served as a starting point for the methodological construction.

Based on the literature review, it can be assumed that stress-inducing circumstances are different for the individual with disability, when compared with the rest of the population. So, the way this individual will handle stress may also be different. Iwasaki and Mannell

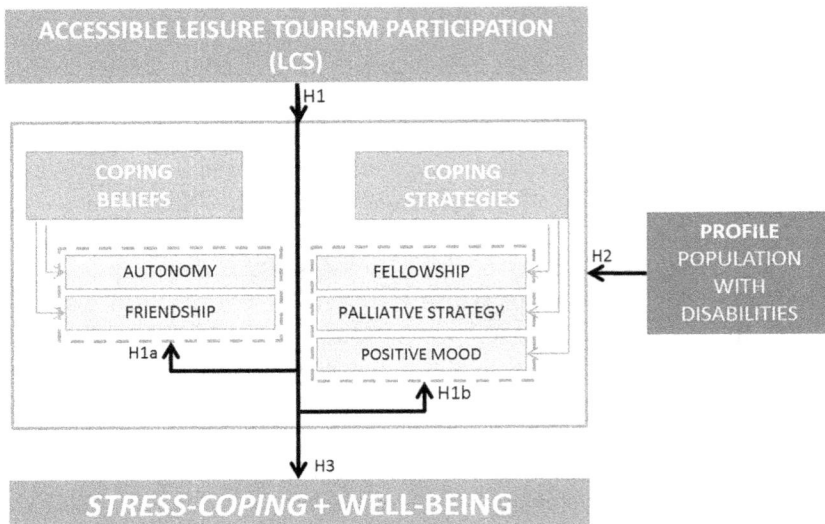

Figure 1. Conceptual model and research hypotheses.

(2000) proposed a hierarchical sub-dimension analysis of stress-coping through leisure, assessed through the Leisure Coping Scales (LCS). Its applicability to the population with disability is the first aim of the present study.

The benefits resulting from participation in accessible leisure tourism will be subject of coping beliefs and strategies, as also suggested and empirically confirmed by Iwasaki and Mannell (2000). Knowing that coping beliefs through leisure provide feelings of autonomy and the promotion of friendship and that coping strategies through leisure promote fellowship, palliative strategies and positive mood, it would be reasonable to think that both beliefs and strategies could have an impact on the stress-coping of individuals with disabilities.

Additionally, this positive effect or feeling of 'inner power' provided by participation in accessible leisure tourism activities leads to the possibility of considering tourism as an effective stress-coping resource for people with disabilities. Thus, it is important to understand the intrinsic and distinctive features of this special population and to understand how tourism can lead to overall well-being (Brown & Hall, 2005; Coleman & IsoAhola,1993; IsoAhola & Park, 1996; Mannell & Kleiber, 1997; Mueller & Kaufmann, 2001). In this case, we suggest the following hypotheses:

H1- Participation in accessible leisure tourism activities positively affects various biopsychosocial stress-coping dimensions of the individual with physical and sensory disability:

H1a – Participation in accessible leisure tourism activities has a positive effect on the coping of people with physical and sensory disabilities, through coping beliefs;

H1b - Participation in accessible leisure tourism activities has a positive effect on the coping of people with physical and sensory disabilities, through the development of coping strategies;

H2- Coping beliefs and strategies through tourism vary, depending on the type of disability and the socio-demographic and economic profile of individuals;

H3- Stress-coping through tourism rebalances and harmonizes the resources of people with physical and sensory disability, positively influencing their overall well-being.

This conceptual model and hypotheses suggested will be validated with data collected from an empirical study, as explained in the following section.

Methodology

We opted for a multidimensional and interdisciplinary strategy, in order to understand the beliefs and coping strategies of individuals with disabilities in a specific leisure context – tourism – trying to simultaneously measure the impact of leisure activities in the context of accessible tourism in the control and management of stress for these individuals. Tourism and especially stress-coping are social phenomena, as they imply contact, interaction and social support (Coleman & IsoAhola, 1993; Hood & Carruthers, 2007; IsoAhola & Park, 1996; Iwasaki, 2001; Iwasaki & Mannell, 2000; Kleiber et al., 2002). Simultaneously, they may be considered individual phenomena, requiring intimate psychological processes, making their study most complex (Jennings & Nickerson, 2006; Kastenholz, Eusébio, Figueiredo, & Lima, 2012). A full understanding of this reality requires both introspection and interpretation, through a phenomenological approach, as suggested by Cohen (1979) and Elands and Lengkeek (2000).

Concerning research design, a transversal, correlational and quantitative study method was used, as detailed next.

Sample

The survey took place in Portugal, over a period of 4 months, from June to September 2013, in person or electronically, according to the convenience of the respondents (people with mobility, visual and hearing impairments). This application period was chosen so as to obtain answers in which the tourist experience had been recent. Additionally, respondents were asked to indicate how long ago the last tourist experience had been, and all those who indicated 'more than one year' were excluded. A 'snowball' sampling technique was used, consisting of the initial contact with some elements of the target population, who indicated other possible participants with the same characteristics or characteristics required by the study, thereby fostering networks of formal and informal contacts relevant to this study.

A total sample of responses from 306 individuals with physical and sensory disability was obtained. This sample consists mainly of individuals between 25 and 44 years of age (60.6%), distributed almost equally by gender (56.4% male, 43.6% female). Out of these respondents, 66.2% present a motor disability, 15.7% a hearing disability and 14.4% a visual disability, and only 3.6% have another sort of incapacity, a category that includes the combination of any of the above (see Figure 2).

When asked about the severity of their disability, survey participants state they have a serious problem (83.7%) and only a few consider that their disability is not a problem (2.9%) (Table 1). This information collection was inspired by the International Classification of Functioning, Disabilities and Health (ICF), whose codes require the use of one or more qualifiers (Functions and Body Structures, Activities and Participation and Environmental Factors) that indicate the magnitude of health or the severity of disability:

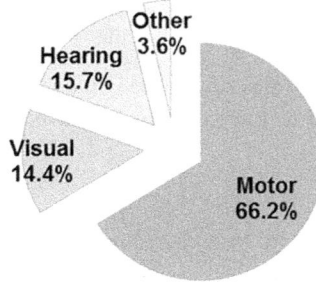

Figure 2. Type of disability.

Table 1. Severity of disability.

Severity of disability	Frequency (No.)	Percentage (%)
Not a problem	9	2.9
Slight problem	2	0.7
Moderate problem	3	1.0
Serious problem	256	83.7
Severe problem	20	6.5
Did not answer	16	5.2
Total	306	100.0

(i) .0 Not a problem (none, absent, insignificant); (ii) .1 Slight problem (light, small); (iii) .2 Moderate problem (medium, regular); (iv) .3 Serious problem (large, extreme); (v) .4 Severe problem (total); (vi) .8 Not specified; (vii) .9 Not applicable (WHO, 2004). As might be expected, 63.2% need some kind of mobility aid, and out of those, 41.7% refer to a wheelchair (manual or electric) as the main technical assistance, followed by crutches (8.1%) and guide dogs for assistance (2.3%).

As far as socio-demographics are concerned (see also summary in Table 2), the respondents' family situation is shaped by most being unmarried (58.9%), while 33.0% are married or have a partner, with a large majority not having any children (71.4%). Interestingly, 44.6% of the respondents are employed and only 14.3% are unemployed (see Figure 3), with the majority showing a high-educational level (39.6% with a higher education degree and 37.6% having completed secondary education), contrary to what was found in previous research concerning this target population in Portugal (Casas, 2005; Centro de Reabilitação Profissional de Gaia, 2007; Gonçalves, 2003).

Table 2. Summary of socio-demographic profile of the sample.

	Variables	Mode
Socio-demographic characteristics	Age group	[25–44] (N = 186)
	Gender	Male (N = 172)
	Type of disability	Motor (N = 202)
	Severity of disability	Serious problem (N = 256)
	Family situation/Marital status	Single (N = 179)
	Parenthood	No children (N = 217)
	Professional situation	Employed (N = 137)
	Qualifications	Higher education (N = 120)
	Income	[€240–€480] (N = 60)

%

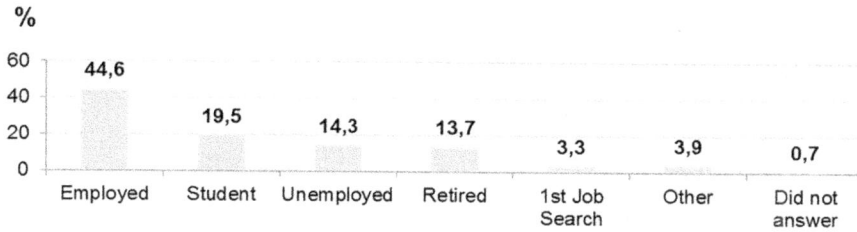

Figure 3. Employment status.

Finally, discretionary monthly income is mostly under €960, and it should be stressed that 19.6% of respondents earn between €240 and €480, 17.3% from €480 to €720, 13.7% from €720 to €960, and 11.1% less than €240.

Considering that the present research focuses on tourism, it was also relevant to understand tourism preferences of the individuals in the study. Thus, 44% of respondents showed a preference for sun and sea activities during their holidays, followed by cultural and nature tourism activities, with respectively 16% and 15% of the preferences. Results also highlight the fact that the level of satisfaction with these activities is quite high, ranging from satisfied (27%) to very satisfied (33%), considering a single-item, 5-point Likert-type scale (1, 'Not satisfied, at all'; 5, 'Very satisfied'). Finally, it is noteworthy that respondents refer to the habit of travelling mostly with family and friends (80%).

Measures

The questionnaire consisted of the adaptation of two scales already tested and validated by Iwasaki and Mannell (2000) and translated into Portuguese by Santos, Ribeiro, and Guimarães (2003) – the Leisure Coping Beliefs Scale (LCBS) and Leisure Coping Strategies Scale (LCSS), introducing two innovations: its application to people with physical and sensory disability in the context of leisure tourism. These scales included, essentially, closed-ended questions, i.e. questions demanding answer choices to be chosen by respondents, depending on their situation, opinion or preferences, particularly measuring behaviour choices and emotions through Likert scale graduations (1, 'Disagree'; 5, 'Strongly Agree').

Based on Iwasaki and Mannell's (2000) model, the adequacy and adjustment of scales applied was initially inspected firstly by calculating the Spearman correlation coefficient between the various stress-coping dimensions, and then the Cronbach's Alpha coefficients.

In this research, the adapted assessment instruments were first validated as to their adjustment and legitimacy. Secondly, results of the adapted scales, appropriate to the context and population in study, were analysed. So, we started with a confirmatory factor analysis rather than with an exploratory factor analysis as the scales adopted were an adjustment of the original LCS, assessing the constructs of the proposed interpretation model. Some variables did not show satisfactory values, requiring the elimination of some items of the original structures of the LCS, resulting in new scales adjusted to accessible leisure tourism, labelled here Leisure Coping Scales in the context of Accessible Leisure Tourism (LCS-ALT – see Appendix I). The results obtained demonstrate the feasibility of the biopsychosocial stress-coping dimensions suggested in the LCS-ALT for a subsequent analysis of the proposed model, contributing to a better understanding of the phenomena under study.

Procedure and data analysis

Descriptive statistics permit sample characterization and specification of general information about biopsychosocial stress-coping dimensions, hierarchically organized by beliefs and strategies within leisure tourism for people with disabilities. Inter-relationships between different variables influencing these coping mechanisms (beliefs and strategies) were assessed by correlational analysis and variable association tests (correlations). Finally, the association between types of disability and socio-demographic characteristics of respondents and the coping beliefs and strategies was checked using bivariate analysis, namely the non-parametric Mann–Whitney U test (when variables did not meet the assumptions of the *t-test*).

Descriptive statistics, normality tests and bivariate analyses were performed using SPSS software, while the confirmatory factor analysis was achieved with LISREL computer software. In short, analysis of the data collected through the questionnaire-based survey helped validate hypotheses H1, H2 and H3.

Results and discussion

The statistical analysis allowed us to understand the relationships suggested in the research model and respective hypotheses, firstly through univariate data analysis by determining absolute frequencies and average values, then through a correlation analysis between constructs, culminating in bivariate analysis, through inferential tests (Mann–Whitney U test and Kruskal–Wallis).

Note that results are generally presented for the whole sample ($N = 306$) and, when appropriate to the research objectives, separated according to three subsamples, corresponding to the three types of disability analysed: motor, hearing and visual.

Leisure coping scales in the context of accessible leisure tourism

Following Iwasaki and Mannell (2000), the Leisure Coping Beliefs Scale, **LCBS-ALT**, may be split into two relevant coping dimensions: **Autonomy** and **Friendship**. Autonomy can be divided into the sub-dimensions of Self-Determination and Empowerment, while Friendship includes the sub-dimensions of Emotional Support, Self-Esteem, Tangible Aid and Information Support.

In a global perspective, descriptive results indicate mean, median and mode values mostly above 3, except for item 29, 'I lack emotional support from my travel companions', averaging 1.91 (SD = 1.16), and median and mode averaging 1 point, but this is an inverted item. Moreover, item 11 stands out: 'what I do within tourism allows me to feel good about myself', with the highest mean (4.40, SD = 0.72), mode (5) and median (5) values.

Results for the sub-dimensions (Self-Determination, Empowerment for Autonomy; Emotional Support, Self-Esteem, Tangible Aid and Information Support for Friendship) are detailed next.

Regarding **Self-determination**, there is a mean of 4.33 (SD = 0.81), which is above average for item 2: 'tourism provides opportunities to regain a sense of freedom'. In the **Empowerment** sub-dimension, the above-cited item 11 is emphasized, but items 4 ('I gain feelings of personal control in tourism') and 19 ('things I do within tourism help me to gain confidence') also have a mean around 4 points. In the case of **Emotional Support**, we highlight item 28 ('I feel emotionally supported by my travel companions'), with a mean of 4.05 (SD = 0.93). Considering **Self-Esteem**, we highlight the top results obtained in every item, emphasizing item 1: 'my travel companions help me feel good about myself' (mean 4.29, SD = 0 90), item 10: 'my travel companions hold me in high esteem' (mean 4.26, SD = 0.83), and item 21:'I am respected by my travel companions' (mean 4.39; SD = 0.80). In **Tangible Aid**, item 25 ('most of my travel companions are happy to take care of my house (apartment), children or pets when I am away') stands out negatively, with a mean of only 2.83 (SD = 1.26). This result, as mentioned in the study of Santos et al. (2003, p. 444), using the same scale, 'may be related to the possible lack of application to the Portuguese context', since most respondents show disability situations that cause physical dependence on others, which explains the fact that many of them live with their families, not having their own residence or pets. Finally, under **Information Support**, all items reach medians above 4 points, except for item 3, 'my travel companions assist me in deciding what to do', with a mean of 3.38 (SD = 1.07).

In short, it appears that both large dimensions have very high degrees of agreement, while the **Empowerment** sub-dimension within the larger dimension **Autonomy** stands out with item 11 (feeling good about oneself through tourism), and sub-dimension **Self-Esteem** standing out in the dimension **Friendship**, with the globally most significant average values.

As far as leisure coping strategies are concerned, the **LCSS-ALT** measures three major dimensions: **Companionship, Palliative Strategy** and **Positive Mood**. Results show that all items reach a median of 3 and 4 points, which means that the majority of the respondents 'agree moderately' (3) or 'strongly' (4) with the statements related to coping strategies through leisure tourism, as is observable in Table 3.

Within the dimension **Companionship**, item 18 stands out: 'one of my strategies to deal with stress was participation in social tourism' (mean = 3.23, SD = 1.23); lower than the other items, with modes of 4 points and means above 3.30, in particular item 5, 'socializing in tourism was a means of managing stress' (mean = 3.51, SD = 1.18). In the **Palliative Strategy** dimension, item 3 – 'I engaged in a tourism activity to temporarily

Table 3. Descriptive results of the dimensions and sub-dimensions of LCS-ALT.

		Mean	Median	Standard deviation (SD)	Min.	Max.	Skewness	Kurtosis
LCBS-ALT		3.85	3.86	0.53	2.17	5.00	−0.37	−0.02
Autonomy	Self-determination	4.10	4.00	0.78	1.00	5.00	−0.99	0.98
	Empowerment	3.98	4.00	0.70	1.33	5.00	−0.70	0.79
Friendship	Emotional support	3.28	3.33	0.54	1.67	5.00	0.09	0.99
	Self-esteem	4.23	4.25	0.68	1.50	5.00	−0.91	0.69
	Tangible aid	3.66	3.69	0.82	1.25	5.00	−0.34	−0.30
	Information support	3.74	3.75	0.71	1.25	5.00	−0.51	0.32
LCSS-ALT		3.51	3.57	0.72	1.00	5.00	−0.33	0.04
Companionship		3.38	3.50	0.99	1.00	5.00	−0.45	−0.30
Palliative strategy		3.21	3.33	0.82	1.00	5.00	−0.26	−0.05
Positive mood		4.09	4.00	0.67	1.00	5.00	−0.73	1.17

get away from the problem' – stands out with lower agreement values. Items 4 – 'escape through tourism was a way of coping with stress' – and 9 – 'tourism was an important means of keeping myself busy' show relatively high agreement values. Finally, items associated with **Positive Mood** present the highest mode (5 points) as well as the higher mean values, above 4.20, more specifically, items 6 – 'I gained a positive feeling from tourism' (mean of 4.20, SD = 0.89) and 10 – 'I maintained a good mood within tourism' (mean of 4.28, SD = 0.79).

In general, the dimension that gets lowest agreement values is the descriptive **Palliative strategy,** with means that vary from 2.50 to 3.50, followed by **Companionship,** averaging from 3.00 to 3.50 approximately, as compared with more positive evaluations of the dimension **Positive Mood** with averages above 3.50.

So, taking some adjustments into account, we worked with an instrument named the LCS-ALT, as shown in Appendix 1. The adaptation of the LCS for the particular context of accessible leisure tourism, considering the special features of a population with motor and sensory disability, proved to be an appropriate procedure for the objectives of this investigation.

Therefore, results of the application of LCS-ALT allow us to draw conclusions about the impacts or benefits of accessible leisure tourism in the management and control of stressful situations, as suggested in hypothesis H1 (Participation in accessible leisure tourism positively affects various biopsychosocial stress-coping dimensions of the individual with physical and sensory disability). Considering both H1a and H1b, a 'positive effect' is assumed when agreement levels are equal to or greater than 2 points, based on the 5-point Likert scale used. Results obtained by descriptive analysis thus confirm hypotheses H1a and H1b, showing that the average values (mean, median and mode) for both scales, LCBS-ALT and LCSS-ALT, are in general above 3 points, proving that participation in accessible leisure tourism activities has a positive effect on coping for people with disabilities, either through the beliefs associated with such participation, or through strategies developed during participation. Similar to other studies in social tourism, disabled people revealed a strong interest in leisure tourism activities, with these activities being considered of utmost importance for self-development (socially, physically and intellectually), self-esteem (becoming proficient in challenge management), life satisfaction and overall quality of life (Kastenholz, Eusébio, & Figueiredo, 2015). The above-mentioned authors argue that these effects are particularly relevant for supporting social tourism policies yielding social inclusion of this underprivileged group of individuals.

For testing the hypothesis regarding coping beliefs and strategies through accessible leisure tourism, correlations with Spearman's coefficient (ρ) were calculated.

LCBS-ALT shows very strong association values with the sub-dimensions **Information Support** ($\rho = 0.83$), **Self-Esteem** ($\rho = 0.82$), **Empowerment** ($\rho = 0.79$), and **Tangible Aid** ($\rho = 0.77$) (see Table 4). These sub-dimensions referring to the same main construct of Coping through Leisure may sometimes be associated, as can be seen through statistical correlational tests (see Table 4). For example, Self-Determination is closely related to Empowerment ($\rho = 0.70$), both belonging to the same sub-dimension 'Autonomy'. Self-Esteem is also correlated with Tangible Aid ($\rho = 0.66$) and Information Support ($\rho = 0.67$), although these last two relations are not as strong as that previously reported.

Results thus highlight Empowerment within Autonomy and Self-Esteem, Tangible Aid and Information Support in the context of Friendship, making it clear that these are the

Table 4. Correlations between LCBS-ALT dimensions and sub-dimensions.

			Autonomy		Friendship			
Spearman coefficient		LCBS-ALT	Self-determination	*Empowerment*	Emotional support	Self-esteem	Tangible aid	Information support
ρ	LCSB-ALT	1.000	0.570**	0.789**	0.540**	0.821**	0.769**	0.832**
	AUTONOMY Self-determination	0.570**	1.000	0.700**	0.147*	0.350**	0.233**	0.291**
	empowerment	0.789**	0.700**	1.000	0.266**	0.529**	0.408**	0.503**
	FRIENDSHIP Emotional support	0.540**	0.147*	0.266**	1.000	0.421**	0.339**	0.500**
	Self-esteem	0.821**	0.350**	0.529**	0.421**	1.000	0.664**	0.671**
	Tangible aid	0.769**	0.233**	0.408**	0.339**	0.664**	1.000	0.658**
	Information support	0.832**	0.291**	0.503**	0.500**	0.671**	0.658**	1.000

** Correlation is significant at 0.01 (2-tailed).
* Correlation is significant at 0.05 (2-tailed).

biopsychosocial dimensions which are most associated with beliefs which individuals with disabilities hold when participating in accessible tourism.

Also, for the domain of **LCSS-ALT**, all dimensions are highly correlated (Table 5). Interestingly, **Companionship** and **Palliative Strategy** show a robust association in statistical terms, with $\rho = 0.73$. Results suggest that when people with disabilities participate in tourism activities, they develop coping strategies that are perfectly connected to each other, including Companionship, Palliative and Positive Mood strategies.

Correlational analysis of both scales (LCBS-ALT and LCSS-ALT) confirm the validity of the adjustments made, with strong associations between multiple selected items and between different dimensions and sub-dimensions assessing stress-coping.

Specifically, it was possible to confirm hypothesis H1 and thereby specify additional benefits of participation in leisure tourism for individuals with a disability, since tourism participation showed a clearly positive effect on all biopsychosocial dimensions analysed in the LCS (regarding both beliefs and strategies), considering univariate and cor-relational analysis results. Of note are the results of the correlation analysis, indicating a strong and robust relationship between all sub-dimensions of LCBS-ALT, except for slightly weaker correlations for the Self-Determination and Emotional Support sub-dimensions and all dimensions of LCSS-ALT, only slightly weaker in the case of Palliative Strategy, suggesting the reinforcement of its measurement (e.g. by increasing the number of items related to each).

The results described above and Iwasaki and Mannell's (2000) theoretical framework allow us to conclude that accessible leisure activities positively influence coping beliefs and strategies (H1a and H1b), which in turn positively affect the various biopsychosocial dimensions of individuals with disabilities (H1). As suggested by Bos, McCabe, and Johnson (2015), experiential learning occurring in a holiday environment stands out in the acquisition of knowledge and skills. Thus, the capacity of coping may be learnt through leisure tourism experiences, which should be accessible to as many people as possible. Given that this group of individuals still reveals serious financial and societal constraints, and that social tourism represents millions of people and is one of the most relevant and promising niches (Bélanger & Jolin, 2011), the implications of these findings for social tourism policies are undeniable.

On the whole, these results reinforce the validity of the LCS-ALT measurement instrument proposed here, and also confirm the significance of tourism for the disabled population in the domain of stress-coping. This suggests a call for corresponding social tourism action yielding the health and social inclusion benefits so urgently needed by this oft-neglected and clearly disadvantaged population group.

Besides, bearing in mind that the focus of this research is a population with very specific intrinsic characteristics, it is imperative to assess the variation of coping beliefs and

Table 5. Correlation between LCSS-ALT dimensions.

Spearman coefficient		LCSS-ALT	Companionship	Palliative strategy	Positive mood
ρ	LCSS-ALT	1.000	0.902**	0.912**	0.712**
	Companionship	0.902**	1.000	0.730**	0.540**
	Palliative strategy	0.912**	0.730**	1.000	0.508**
	Positive mood	0.712**	0.540**	0.508**	1.000

** Correlation is significant at 0.01 (2-tailed). `
* Correlation is significant at 0.05 (2-tailed).

strategies depending on the type of disability and the socio-demographic and economic profile of the case study population. Thus, through bivariate analyses relating dimensions and sub-dimensions of LCBS-ALT and LCSS-ALT with the variables (a) type of disability, (b) gender, (c) marital status, (d) parenthood, (e) qualifications and (f) income, it was possible to confirm hypothesis H2.

Factors conditioning the impact of coping beliefs and strategies through tourism for persons with disability

In order to assess the impact of a series of profiling variables on responses to the LCBS-ALT and LCSS-ALT variables, we initially tested ANOVA assumptions of normal distribution and homoscedasticity. As none of the assumptions were verified, we opted for the non-parametric alternatives: Kruskall–Walllis and Mann–Whitney U test for verifying the relationship between a dichotomous variable (if two categories, Mann–Whitney U, If more, Kruskall–Wallis) and the scale variables, i.e. to compare two or more independent samples regarding the ordinal scale variable. For purposes of interpretation, only results with a 5% significance level are considered.

In the following sub-sections, Mann–Whitney U and Kruskall–Wallis test values are presented, with significance values and, finally, global means of ranks revealing the level of agreement (in brackets), for analysing the relationship between the coping belief and strategy scales and: (a) type of disability, (b) gender, (c) marital status, (d) parenthood, (e) qualifications and (f) income.

(a) Type of disability
The Kruskal–Wallis test showed that LCBS-ALT is influenced by respondents' type of disability ($X_{KW}2$ (2) = 21.45; $p = 0.00$), in its sub-dimensions of Information Support ($X_{KW}2$ (2) = 8.72; $p = 0.01$), Self-Determination ($X_{KW}2$ (2) = 13.13; $p = 0.00$), Empowerment ($X_{KW}2$ (2) = 14.43; $p = 0.00$), Self-Esteem ($X_{KW}2$ (2) = 30.53; $p = 0.00$) and Tangible Aid ($X_{KW}2$ (2) = 38.94; $p = 0.00$).

The Dunn test for non-parametric pairwise multiple comparisons based on rank sums permitted the identification of the exact position of significant differences (Marôco, 2010). The Dunn tests for **LCBS-ALT** as a whole indicate significant statistical differences ($p <$ 0.05) between hearing and visual impairment and between hearing and motor impairment. No significant differences between visual and motor impairment were found. Specifically, for both **Autonomy** sub-dimensions (Self-Determination and Empowerment) individuals with motor disability show a higher Mean Rank value (159.14 for **Self-Determination**, 159.40 for **Empowerment**), meaning that they are the ones who agree most with achieving higher levels of autonomy through accessible leisure tourism, when compared to those with hearing impairment. Considering the sub-dimensions Emotional Support, Self-esteem, Tangible Aid and Information Support, all within the **Friendship** dimension, the Dunn test shows that there are differences in the sub-dimension **Self-Esteem**, where individuals with a motor disability show higher levels of agreement (162.37), when compared to those with hearing impairment. In the sub-dimensions **Tangible Aid** and **Information Support**, people with visual impairment show higher Mean Rank values (162.27 for Tangible Aid, 160.69 for Information Support) when compared to those with hearing impairment (Mean rank of 77.76 and

115.25, respectively), revealing higher levels of agreement to the corresponding support received from travel mates by those with visual impairment.

LCSS-ALT is also influenced by disability typology ($X_{KW}2$ (2) = 18,43, $p < 0.001$), but Dunn tests only present significant statistical differences ($p < 0.05$) between motor and hearing impairments in the sub-dimension **Positive Mood**. As indicated by the mean rank values, individuals with motor impairment are those showing stronger agreement with the positive effects associated with mood (161.18 versus 159.10).

(b) Gender

As far as gender is concerned, it was found that neither **LCBS-ALT** nor its dimensions show any significant statistical difference. Thus, gender apparently does not impact leisure coping beliefs.

Leisure coping strategies assessed by **LCSS-ALT**, on the contrary, reveal higher mean values for women ($M = 3.61$; SD = 0.69) compared with their male peers ($M = 3.43$; SD = 0.74) and differences were statistically significant ($U = 9894.00$; MW = 24772.00; $p = 0.04$). Results demonstrated significant differences for the dimensions of **Companionship** ($U = 9546.50$; $W = 24424.50$; $p = 0.01$) and **Palliative Strategy** ($U = 9932.00$; $W = 24810.00$; $p = 0.05$), where women apparently perceive more benefits in stress-coping through accessible leisure tourism than men.

(c) Marital status

The variable 'marital status' only affects the **Information Support** sub-dimension in **LCBS-ALT** ($X_{KW}2$ (3) = 10.03; $p = 0.02$). Dunn tests reveal significant statistical differences ($p < 0.05$) between married and single respondents, with single individuals agreeing the most with the effects provided by the Information Support sub-dimension (Mean Rank = 163.42 versus 130.27 for married individuals). Single individuals with disabilities may lack the consistent support of a partner, therefore feeling the need of information support more intensely when considering stress reduction through leisure tourism.

Likewise, **LCSS-ALT** is also influenced by the marital status of respondents: $X_{KW}2$ (3) = 25.34; $p < 0.001$. Results pointed out statistically significant differences ($p < 0.001$) in LCSS-ALT between the categories 'single' and 'married', with unmarried individuals again presenting a higher Mean Rank value (168.35) when compared to married respondents (111.99). They correspondingly agree most with the effects associated with coping strategies through leisure tourism, particularly in terms of **Companionship** (Mean Rank = 166.99) and **Palliative Strategy** (Mean Rank = 169.68).

(d) Parenthood

On average, **LCBS-ALT** shows statistically significant differences ($U = 7297.00$; $W = 11125.00$; $p = 0.00$) between respondents without children ($M = 3.91$; SD = 0.51) and those with children ($M = 2.69$; SD = 0.56), in its sub-dimensions: **Empowerment** ($U = 7721.50$; $W = 11549.50$; $p = 0.05$). **Self-Esteem** ($U = 8103 \ 50$; $W = 12068.50$; $p = 0.00$) and **Information Support** ($U = 6451.50$; $W = 10279.50$; $p = 0.01$) all revealing the same direction. That is, those without children perceive more benefits from accessible tourism in the mentioned dimensions, particularly regarding enhancement of autonomy as well as the role of travel mates presenting information support to these respondents.

Similarly, **LCSS-ALT** reveals higher mean values among respondents without children ($M = 3.60$; $SD = 0.70$) compared to respondents with children ($M = 3.28$; $SD = 0.72$) and differences were statistically significant ($U = 6935.50$; $W = 10763.50$; $p < 0.001$), especially in the dimensions of **Companionship** and **Palliative Strategy** ($U = 7042.00$; $W = 10870.00$; $p < 0.001$ and $U = 6909.50$; $W = 10737.50$; $p < 0.001$, respectively), with no relevant difference visible for the Positive Mood dimension.

Results of this analysis may be related to the previous results regarding marital status, with an active role in both marriage and parenthood suggesting a distinct general structure of daily life together with the partner and rest of close family. This probably also impacts on daily leisure and the consequent increased meaning of leisure tourism for those lacking such a daily family structure.

(e) Qualifications
No statistically significant differences in terms of education level could be found either for **LCBS-ALT** or for **LCSS-ALT** and respective sub-dimensions.

(f) Income
Income influences **LCBS-ALT** ($X_{KW}2$ $(8) = 18.34$. $p = 0.02$) i.e. in at least one of the salary levels the scale distribution is statistically different in a significant way. Income also significantly influences the **Information Support** and **Empowerment** sub-dimensions $X_{KW}2$ $(8) = 19.30$. $p = 0.01$ and $X_{KW}2$ $(8) = 22.74$. $p = 0.00$, respectively.

Dunn tests confirm the statistically significant inequalities ($p < 0.05$) in LCBS-ALT (globally) and show specifically that individuals with the lowest level of income (those within wage category 'Less than 240 €') present a higher mean rank value for the whole LCBS-ALT ($p = 158.41$) and for the two mentioned sub-dimensions (Mean Rank = 162.69 for **Information Support**; Mean Rank = 157.74 for **Empowerment**). This means that respondents with lower incomes agree the most with the general beneficial effects suggested by LCBS-ALT, particularly regarding the effects of the sub-dimensions **Information Support** and **Empowerment**.

LCSS-ALT is also influenced by income $X_{KW}2$ $(8) = 24.76$. $p = 0.00$, particularly for the **Companionship** and **Palliative Strategy** dimensions. Results show that individuals with the lowest incomes (<240 €) are those with higher mean rank values considering the whole LCSS-ALT (158.41), as well as the Companionship (161.99) and Palliative Strategy (169.57) dimensions. These individuals again show most agreement with the general effects of coping strategies through leisure tourism reflected in LCSS-ALT and specifically regarding its sub-dimensions Companionship and Palliative Strategy.

Zenko and Sardi (2014) claim the usual research approach to people with special needs in the context of social tourism is too one-sided and recommend a more substantial and holistic methodology to uncover this population groups' particular needs, benefits derived from and attitudes towards participation in tourism. Accordingly, results showed that coping beliefs of individuals with disabilities in accessible leisure tourism contexts vary according to the type of disability, marital status, parenthood and income, while neither gender nor educational background influence this category (coping beliefs) as a whole.

As far as the sub-dimensions of Autonomy (Self-Determination and Empowerment) are concerned, which vary between hearing and motor impairment, it was found that individuals with physical disability were those who agreed most with the positive effects suggested in this dimension. These results are justified, as people with mobility difficulties tend to be subject to greater physical accessibility constraints and dependence on others to perform daily life activities, unlike deaf people, who face mainly communication barriers. Hence, Autonomy is a factor of greater importance for individuals with physical disabilities. Considering the Friendship sub-dimension, it is important to highlight Self-esteem, Tangible Aid and Information Support, with variations among individuals with hearing and motor impairments and between individuals with hearing and visual impairment. Self-esteem is more valued by people with motor disability, and the Tangible Aid and Information Support is more relevant for individuals with visual impairment. This variation is also understandable, as individuals with motor impairment will be those, whose disability is more visible, and therefore tend to be the ones who need to work on Self-Esteem the most. In turn, Tangible Aid and Information Support will be particularly relevant for blind people, as the simple impossibility of assessing visual cues from surroundings in an unfamiliar environment may trigger a stressful situation.

Marital status only influences the sub-dimension Information Support among single versus married people and is more significant for single individuals, while married individuals apparently do not value this aspect so much or are less concerned about it, possibly due to confidence in their partner. Similarly, and probably related to marital status, parenthood (having children or not) affects coping beliefs in general and, in particular, the Empowerment, Self-Esteem and Information Support sub-dimensions, since individuals without children may be more concerned with these leisure coping beliefs, as they are less integrated in a family structure and therefore more exposed to corresponding travel challenges.

Finally, considering income, it appears that those with lower incomes (less than 240€ per month) are the ones that agree the most with the effects of coping beliefs in the accessible leisure tourism context, especially regarding the Empowerment and Information Support sub-dimensions. This result is consistent with other studies suggesting that people from lower social classes would have greater willingness to participate in and be likely to derive greater benefits from leisure activities (Iwasaki, 2006; Mannell & Kleiber, 1997). This should be even more true for leisure in a tourism context, which is less available and thus more valued by poorer population groups, as found for low-income families (Lima, 2017) and also for many Portuguese individuals with disability, subject to social exclusion in diverse life spheres (Kastenholz et al., 2015).

Results also reveal that coping *strategies* for people with disabilities in the accessible leisure tourism context, as opposed to coping *beliefs*, vary within all defined socio-demographic categories, suggesting more heterogeneity of responses to the scales reporting actual travel challenges and how people coped with them in contrast to more general beliefs and attitudes regarding travel and tourism.

In general, results suggest that the proposed stress-coping model is valid in the leisure tourism context for people with physical and sensory disability. Therefore, also confirming other studies, there is strong evidence for assuming that individuals with disabilities develop coping beliefs and strategies through leisure tourism, reorganizing and reinforcing their internal resources, which allows them to handle, control and manage stressful situations, with obvious positive impacts on their overall well-being (Iwasaki & Schneider, 2003; Iwasaki, MacTavish, & MacKay, 2005; Zuzanek et al., 1998). These results also suggest the close link between accessible leisure tourism and social tourism, since tourism was identified as an important tool for health, rehabilitation and social inclusion of persons with a disability, a generally marginalized group (Kastenholz et al., 2015). In this context, tourism may boost health benefits within combined tourism-rehabilitation programmes, as rehabilitation aims to enhance and restore functional ability and quality of life to those with impairments or disabilities, maximizing independence in activities of daily living (Kleiber et al., 2008; Mactavish & Iwasaki, 2005). Also, Carneiro et al. (2013) stress the potential of social tourism for generating important social benefits for persons with disabilities. Several other studies corroborate the potential of social tourism for enhancing social inclusion, health and well-being, as in the case of social tourism programmes directed to the elderly (Eusébio, Carneiro, Kastenholz, & Alvelos, 2016, 2017) or to young people (Foster & McCabe, 2015), with positive effects shown between tourist experiences and individual as well as social outcomes. Thus, considering both the social and therapeutic perspectives, one may suggest personal and social well-being as a final goal, summarizing the interpretations given to coping beliefs or strategies used by people with disabilities, within a social and leisure tourism context. Consequently, hypothesis H3 is confirmed, since stress-coping rebalances and harmonizes resources of people with physical and sensory disability, positively influencing their overall well-being.

Conclusions

Briefly, our results from both the literature review and empirical research respond to the proposed conceptual model and research objectives, as follows:

1. The LCS can be used in the context of accessible leisure tourism with the necessary adaptations, serving as a basis for assessing and interpreting the biopsychosocial dimensions of stress-coping developed by people with disabilities.
2. Coping beliefs associated with accessible leisure tourism are strongly related to biopsychosocial dimensions such as Autonomy, especially in its Empowerment sub-dimension, and Friendship, with proven emphasis on Self-Esteem in all types of disability;
3. Coping strategies developed during participation in accessible leisure tourism promote several biopsychosocial dimensions: Companionship, Palliative Strategy and Positive Humour, underlining Positive Mood as the dimension most valued by all individuals, regardless of their type of disability;
4. Coping beliefs and strategies associated with accessible leisure tourism apparently strengthens skills and capabilities of individuals with motor and sensory disability to handle, control and manage stress, supporting the rebalancing and harmonization of their personal and social resources, positively contributing to their health and global well-being.

These final conclusions suggest theoretical and practical implications for social inclusion-enhancing policies, specifically through accessible social tourism programmes, so as to enhance the opportunity of the disabled population of gaining new sources of stress-coping, empowerment and social inclusion through means of accessible leisure tourism. Social tourism initiatives thus seem desirable, increasing disadvantaged groups' participation in tourism activities (Eusébio et al., 2016, 2017; Kastenholz et al., 2015). The theoretical evolution of the concepts involved in this debate as well as empirical evidence is as necessary here as progress of social awareness and corresponding social policies leading to increased accessible leisure tourism opportunities in practice (Kastenholz et al., 2015).

The main theoretical implications of the present study are: (i) clarification of the stress-coping phenomenon in the accessible leisure tourism context, particularly in terms of benefits or positive impacts on the development of biopsychosocial resources for individuals with motor and sensory disability; (ii) identification of social benefits as one of the main benefits of accessible leisure tourism; (iii) recognition of accessible leisure tourism as the preferred coping response for people with physical and sensory disability, through social dimensions, such as enhancement of Friendship and Companionship.

Some practical contributions of the present research project are: (i) enhanced awareness of the universal right to participation in tourism, including people with disabilities; (ii) increased general awareness of the added value and benefits of accessible leisure tourism for social inclusion; (iii) systematization of the biopsychosocial dimensions that are positively influenced by accessible leisure tourism, which could serve as a basis for planning new therapeutic interventions for the rehabilitation of individuals with physical and sensory disability; (iv) understanding accessible leisure tourism as a management model for the development of innovative social tourism products, based on social integration and promotion of health and well-being, in order to improve the destination's differentiation and competitiveness (Kastenholz et al., 2012). Accessible leisure tourism as a social tourism activity was shown to improve not only participants' health but also to promote positive effects for socio-economic development of destinations (Eusébio, et al., 2016; Vinogradova, Larionova, Suslova, Povorina, & Korsunova, 2015). Overall benefits resulting from social tourism are definitely of utmost importance for consequent policy development (Eusébio et al., 2016, 2017; Kastenholz et al., 2015).

Despite the relevance and timeliness of this paper to the progress of knowledge in tourism, some limitations and constraints of this study must be acknowledged. Firstly, the type of population studied here implies sample access constraints, since people with disabilities are a niche population in Portugal, relatively excluded from social life (Kastenholz et al., 2015), with relatively few qualifications, imposing difficulties in accessing a large number of respondents and implementing complex survey instruments with a certain level of abstraction. Secondly, constraints regarding the types of disability made data collection (frequently dependent on the availability of helpful intermediaries) difficult, involving delays in the survey process. Lastly, the variety of disabilities still leads to obstacles related to the need to adapt investigation techniques and tools, ideally requiring adapted research approaches.

Being conscious that limitations may turn into challenges to enhance knowledge and understanding of the investigated theme, we suggest the creation and validation of tools suited to different types of disability, which promote more a robust, consistent and proper assessment of the impact of tourism in relieving stress and consequently in promoting well-being and quality of life for the population with disabilities, in accordance with the heterogeneity of the group of individuals with disabilities (Figueiredo, Eusébio, & Kastenholz, 2012).

The results presented also alert us to the need for further studies to more clearly examine cultural, contextual and disability-based differences of the group of individuals studied here. Moreover, it would be of utmost interest to understand the differences between disabled individuals experiencing leisure tourism in the context of social tourism programmes and the general population. Also, the role of therapeutic intervention policies for individuals with disabilities in the context of rehabilitation deserves further study, so as to better understand and stimulate the potential of concerted action of all those involved in decision-making processes, either in tourism, social or health sectors. According to Diekmann and McCabe (2011), a strong European policy for social tourism could be the answer, through the implementation of an inclusive system, beyond the economic framework mainstream. We subscribe to Minnaert, Maitland, and Miller's (2006, 2009, 2011) contention that social tourism will not provide benefits spontaneously, since welfare agents need to be aware of their role in the process and to know exactly how they may encourage the desired outcomes. Continuous research enhancing our understanding and providing action-relevant orientation in the field is still needed.

Today, accessible tourism organizations depend on state intervention together with the voluntary sector, with its conception of 'social tourism' supported by taxes and/or enabled by charities and volunteers (Walton, 2013, p. 48). However, in order to reach the required level of social tourism development, structural, legislative, institutional and strategic aspects of public decision-making need to be considered (Vinogradova et al., 2015). In this context, La Placa and Corlyon (2014) suggest an interesting bottom-up development approach, through the introduction of social tourism at the local and community level, where local stakeholders and communities collaborate in social tourism partnerships and networks. In conclusion, independently of the political and destination context, this research paves the way to a new area of interest in the field of accessible tourism: the role of accessible leisure tourism in stress-coping of individuals with a physical and sensory disability. Emphasizing the potential of tourism as an effective means to promote social inclusion, health and well-being of individuals with disabilities, we argue that this potential should be implemented for every age group from childhood to old age. Thereby, accessible leisure tourism can contribute to primary prevention, avoidance of stress, anxiety, depression, health problems in general, and to secondary prevention, promoting rehabilitation. All these benefits should be understood as relevant for both the disabled individual's personal well-being and that of society overall.

Disclosure statement

No potential conflict of interest was reported by the authors.

ORCID

Andreia Filipa Antunes Moura 🄳 http://orcid.org/0000-0002-1722-3476

References

Assipova, Z., & Minnaert, L. (2014). Tourists of the world, unite! The interpretation and facilitation of tourism towards the end of the Soviet Union (1962–1990). *Journal of Policy Research in Tourism, Leisure and Events*, 6(3), 215–230.

Bélanger, C. E., & Jolin, L. (2011). The International Organisation of Social Tourism (ISTO) working towards a right to holidays and tourism for all. *Current Issues in Tourism, 14*(5), 475–482.

Bos, L., McCabe, S., & Johnson, S. (2015). Learning never goes on holiday: An exploration of social tourism as a context for experiential learning. *Current Issues in Tourism, 18*(9), 859–875.

Bramston, P., & Fogarty, G. (1995). Measuring stress in the mildly intellectually handicapped: The factorial structure of the subjective stress scale. *Research in Developmental Disabilities, 16*(2), 117–131.

Brown, F., & Hall, D. (2005). *Tourism and welfare: Ethics, responsibility and sustained well-being.* Wallingford: CAB International.

Carneiro, M. J., Eusébio, C., Kastenholz, E., & Alvelos, H. (2013). Motivations to participate in social tourism programmes: A segmentation analysis of the senior market. *Anatolia, 24*(3), 352–366.

Casas, C. L. J. (2005). *Lazer sem barreiras – Guia de turismo adaptado para pessoas com deficiência.* Lisboa: Projecto CAMI 2004.

Centro de Reabilitação Profissional de Gaia, C. R. P. G. (2007). *Mais Qualidade de Vida para as Pessoas com Deficiências e Incapacidades – Uma Estratégia para Portugal.* Gaia: Centro de Reabilitação Profissional de Gaia.

Cohen, E. (1979). A phenomenology of tourist experiences. *Sociology, 13*(2), 179–201. doi:10.1177/003803857901300203

Coleman, D., & IsoAhola, S. E. (1993). Leisure and health: The role of social support and self-determination. *Journal of Leisure Research, 25*(2), 111–128.

Diekmann, A., & McCabe, S. (2011). Systems of social tourism in the European Union: A critical review. *Current Issues in Tourism, 14*(5), 417–430.

Elands, B. H. M., & Lengkeek, J. (2000). *Typical tourists: Research into the theoretical and methodological foundations of a typology of tourism and recreation experiences.* Leiden: Backhuys Publishers.

Eusébio, C., Carneiro, M. J., Kastenholz, E., & Alvelos, H. (2016). The impact of social tourism for seniors on the economic development of tourism destinations. *European Journal of Tourism Research, 12*, 5–24.

Eusébio, C., Carneiro, M. J., Kastenholz, E., & Alvelos, H. (2017). Social tourism programmes for the senior market: A benefit segmentation analysis. *Journal of Tourism and Cultural Change, 15*(1), 59–79. doi:10.1080/14766825.2015.1117093

Figueiredo, E., Eusébio, C., & Kastenholz, E. (2012). How diverse are disabled tourists? A pilot study on accessible leisure tourism experiences in Portugal. *International Journal of Tourism Research, 14*(6), 531–550.

Folkman, S., & Moskowitz, J. T. (2004). Coping pitfalls and promise. *Annual Review of Psychology, 55*, 745–774.

Foster, C., & McCabe, S. (2015). The role of liminality in residential activity camps. *Tourist Studies, 15*(1), 46–64.

Gonçalves, C. (2003). Enquadramento familiar das pessoas com deficiência: Uma análise exploratória dos resultados dos censos 2001. *Estudos Demográficos, 33*, 71–94.

Hood, C., & Carruthers, C. (2007). Enhancing leisure experience and developing resources: The leisure and well-being model. *Therapeutic Recreation Journal. Fourth Quarter* 2007. Retrieved from http://findarticles.com/p/articles/mi_qa3903/is_200901/ai_n31964286/?tag=content;col1

IsoAhola, S. E., & Park, C. J. (1996). Leisure-related social support and self-determination as buffers of stress-illness relationship. *Journal of Leisure Research, 28*(3), 169–187.

Iwasaki, Y. (2001). Testing an optimal matching hypothesis of stress. Coping and health: Leisure and general coping. *Loisir et Société, 24*(1), 163–203.

Iwasaki, Y. (2006). Counteracting stress through leisure coping: A prospective health study. *Psychology. Health & Medicine, 11*(2), 209–220.

Iwasaki, Y., MacTavish, J., & MacKay, K. (2005). Building on strengths and resilience: Leisure as a stress survival strategy. *British Journal of Guidance & Counselling, 33*(1), 81–100. doi:10.1080/03069880412331335894

Iwasaki, Y., & Mannell, R. C. (2000). Hierarchical dimensions of leisure stress coping. *Leisure Sciences*, *22*(3), 163–181.

Iwasaki, Y., & Schneider, I. E. (2003). Leisure, stress and coping: An evolving area of inquiry. *Leisure Sciences*, *25*(2–3), 107–113. doi:10.1080/01490400390211781

Jennings, G., & Nickerson, N. P. (2006). *Quality tourism experiences*. Oxford: Elsevier Butterworth–Heinemann.

Kastenholz, E., Eusébio, C., & Figueiredo, E. (2015). Contributions of tourism to social inclusion of persons with disability. *Society and Disability*, *30*(8), 1259–1281.

Kastenholz, E., Eusébio, C., Figueiredo, E., & Lima, J. (2012). Accessibility as competitive advantage of a tourism destination: The case of Iousã. In Hyde, Ryan & Woodside (Eds.), *Field guide for case study research in tourism, hospitality, and leisure (advances in culture, tourism and hospitality research, volume 6)* (pp. 369–385). Bingley: Emerald.

Kastenholz, E., Eusébio, C., Moura, A., & Figueiredo, E. (2010). Animação turística acessível: do sonho à realidade. In ISCE (Ed.), *Turismo acessível: Estudos e experiências* (pp. 171–192). Lisboa: Edições Pedagogo.

Kleiber, D. A., Hutchinson, S. L., & Williams, R. (2002). Leisure as a resource in transcending negative life events: Self-protection. Self-restoration. And personal transformation. *Leisure Sciences*, *24*(2), 219–235.

Kleiber, D. A., Reel, H. A., & Hutchinson, S. L. (2008). When distress gives way to possibility: The relevance of leisure in adjustment to disability. *Neurorehabilitation*, *23*(4), 321–328.

La Placa, V., & Corlyon, J. (2014). Social tourism and organised capitalism: Research, policy and practice. *Journal of Policy Research in Tourism, Leisure and Events*, *6*(1), 66–79.

Lazarus, R. (1993). Coping with stress of illness. In A. Kaplun (Ed.), *Health promotion and chronic illness: Discovering a new quality of health* (pp. 11–29). Copenhagen: WHO Regional Office for Europe.

Lazarus, R., & Folkman, S. (1984). *Stress appraisal and coping*. New York, NY: Springer.

Lima, J. (2017). Family tourism: The importance of tourism for low-income families. *European Journal of Tourism Research*, *15*, 167–170.

Mactavish, J., & Iwasaki, Y. (2005). Exploring perspectives of individuals with disabilities on stress-coping. *Journal of Rehabilitation*, *71*(1), 20–31.

Mannell, R. C., & Kleiber, D. A. (1997). *A social psychology of leisure*. State College, PA: Venture.

Marôco, J. P. (2010). *Análise Estatística com o PASW Statistics (ex-SPSS)* (1st ed.). Pêro Pinheiro: ReportNumber.

McCabe, S., & Diekmann, A. (2015). The rights to tourism: Reflections on social tourism and human rights. *Tourism Recreation Research*, *40*(2), 194–204.

McCabe, S., Minnaert, L., & Diekmann, A. (2012). *Social tourism in Europe: Theory and practice*. Bristol: Channel View.

Minnaert, L., Maitland, R., & Miller, G. (2006). Social tourism and its ethical foundations. *Tourism, Culture and Communication*, *7*(1), 7–17.

Minnaert, L., Maitland, R., & Miller, G. (2009). Tourism and social policy: The value of social tourism. *Annals of Tourism Research*, *36*(2), 316–334.

Minnaert, L., Maitland, R., & Miller, G. (2011). What is social tourism? *Current Issues in Tourism*, *14*(5), 403–415.

Mueller, H., & Kaufmann, E. L. (2001). Wellness tourism: Market analysis of a special health tourism segment and implications for the hotel industry. *Journal of Vacation Marketing*, *7*(1), 5–17.

Neal, J., Uysal, M., & Sirgy, M. (2007). The effect of tourism service on travellers' quality of life. *Journal of Travel Research*, *46*, 154–163.

Ryan, C. (2002). Equity, management, power sharing and sustainability-issues of the 'new tourism'. *Tourism Management*, *23*(1), 17–26.

Santos, L. R., Ribeiro, J. P., & Guimarães, L. (2003). Estudo de uma escala de crenças e de estratégias de *coping* através do lazer. *Análise Psicológica*, *21*(4), 441–451.

Schneider, I. E., & Iwasaki, Y. (2003). Reflections on leisure stress and coping research. *Leisure Sciences*, *25*(2–3), 301–305. doi:10.1080/01490400390211871

Shaw, G., & Coles, T. (2004). Disability holiday making and the tourism industry in the UK: A preliminary survey. *Tourism Management, 25*(3), 397–403. doi:10.1016/S0261-5177(03)00139-0

Vinogradova, M. V., Larionova, A. A., Suslova, I. A., Povorina, E. V., & Korsunova, N. M. (2015). Development of social tourism: Organizational, institutional, and financial aspects. *Regional and Sectoral Economic Studies, 15*(2), 123–136.

Walton, J. K. (2013). 'Social tourism' in Britain: History and prospects. *Journal of Policy Research in Tourism. Leisure and Events, 5*(1), 46–61.

World Health Organization, W.H.O. (2004). *ICF – international classification of functioning, disability and health.* Geneva: World Health Organization.

Zenko, Z., & Sardi, V. (2014). Systemic thinking for socially responsible innovations in social tourism for people with disabilities. *Kybernetes, 43*(3), 652–666.

Zuzanek, J., Robinson, J. P., & Iwasaki, Y. (1998). The relationships between stress, Health, and physically active leisure as a function of life-cycle. *Leisure Sciences, 20*(4), 253–275.

Appendix 1

LCS-ALT: Leisure Coping Scales in the context of Accessible Leisure Tourism

LCBS-ALT: Leisure Coping Beliefs Scale in the context of Accessible Leisure Tourism

			LCBS-ALT	
		Number	Item
Autonomy	Self-Determination	lcbs2	Tourism provides opportunities to regain a sense of freedom
		lcbs5	Tourism is a self-determined activity for me
	Empowerment	lcbs 4	I gain feelings of personal control in tourism
		lcbs11	What I do within tourism allows me to feel good about myself
		lcbs15	I am able to openly express who I am in my tourism time
		lcbs19	The things I do within tourism help me gain confidence
		lcbs26	My tourism participation enhances my self-concept
		eccl30	Opportunities to express myself in tourism activities enhance my self-concept
Friendship	Emotional support I	lcbs7	My travel companions listen to my private feelings
		lcbs28	I feel emotionally supported by my travel companions
		lcbs29	I lack emotional support from my travel companions
	Self-Esteem	lcbs1	My travel companions help me feel good about myself
		lcbs10	My travel companions hold me in high esteem
		lcbs21	I'm respected by my travel companions
		lcbs24	I feel that I'm valued by my travel companions
	Tangible aid	lcbs12	When I need to borrow something, my travel companions will lend it to me
		lcbs14	If I need extra hands for doing tasks, I can turn to my travel companions
		lcbs16	My travel companions would lend me money if necessary
		eccl25	Most of my travel companions are happy to take care of my house (apartment), children, or pets when I am away
	Information	lcbs3	My travel companions assist me in deciding what to do
	Support	lcbs18	My travel companions give me advice when I am in trouble
		lcbs20	My travel companions often provide me with useful information
		lcbs23	I can talk to my travel companions when I am not sure what to do

LCSS-ALT: Leisure Coping Strategies Scale in the context of Accessible Leisure Tourism

	Number	Item
		LCSS-ALT
Companionship	lcss5	Socializing in tourism was a means of managing stress
	lcss7	I dealt with stress through spending tourism time with my friends
	lcss8	Engaging in social tourism was a stress-coping strategy for me
	lcss18	One of my strategies to deal with stress was participation in social tourism
Palliative strategy	lcss3	I engaged in a tourism activity to temporarily get away from the problem
	lcss4	Escape through tourism was a way of coping with stress
	lcss9	Tourism was an important means of keeping myself busy
	lcss11	Engagement in tourism allowed me to gain a fresh perspective on my problem(s)
	lcss14	By escaping from the problem through tourism, I was able to tackle my problem(s) with renewed energy
	lcss17	I took a brief break through tourism to deal with the stress
Positive mood	lcss1	Tourism helped me feel better
	lcss6	I gained a positive feeling from tourism
	lcss10	I maintained a good mood within tourism
	lcss16	Tourism helped me manage my negative feeling

Accessible tourism experiences: the voice of people with visual disabilities

Eugenia Devile 🅳 and Elisabeth Kastenholz

ABSTRACT

Our research focuses primarily on the understanding of the engagement in tourist activities of people with disabilities. Specifically, we intend to analyze the experience of people with visual impairments, identifying the factors that constrain and the factors that facilitate their decision to travel, seeking to understand how people adapt, negotiate their perceived and real constraints and become active travelers. A qualitative study was undertaken using in-depth interviews, which sought to give voice to people with visual disabilities. Results of a content analysis of the information gathered allow the following main conclusions: The participation in tourist activities by people with visual disabilities results from an ongoing and interactive process, which is shaped by multiple factors, with positive or negative influences, within each individual's very personal context, impairment condition and social environment, with impact on different stages of the process. It is possible to identify patterns of factors influencing the process (constraints and facilitators) as well as of negotiation strategies used for continued participation in tourism. To overcome the barriers they face, people with disabilities rely on negotiation strategies associated with the travel organization process, and on different personal and interpersonal strategies that are systematized in this study

RESUMEN

Nuestra investigación se enfoca principalmente en la comprensión del compromiso en actividades turísticas de las personas con discapacidad. Específicamente, nos proponemos analizar la experiencia de las personas con deficiencia visual, identificando los factores que limitan y los que facilitan su decisión de viajar, buscando comprender cómo se adaptan, negocian sus limitaciones percibidas y reales y se convierten en viajeros activos. Se llevó a cabo un estudio cualitativo utilizando entrevistas en profundidad que buscaban dar voz a las personas con deficiencia visual. Los resultados de un análisis de contenido de la información recogida permite alcanzar las siguientes conclusiones principales: La participación en actividades turísticas de las personas con deficiencia visual es el resultado de un proceso continuo e interactivo que está conformado por múltiples factores, con influencias positivas o negativas, dentro del contexto personal de cada individuo, la discapacidad y el entorno social, con impacto en las diferentes etapas del proceso. Es posible identificar patrones de factores que influyen en el proceso (limitaciones y facilitadores) así como las estrategias de negociación utilizadas para una participación continuada en la actividad turística. Para superar las barreras a las que se enfrentan, las personas con discapacidad se basan en las estrategias de negociación asociadas con el proceso de organización del viaje y en diferentes estrategias personales e interpersonales que son sistematizadas en este estudio.

RÉSUMÉ

La présente recherche a pour objection principal la compréhension de l'engagement des personnes handicapées dans les activités touristiques. Plus précisément, on a l'intention d'analyser l'expérience des personnes ayant une déficience visuelle, d'identifier les facteurs contraignants et les facteurs qui facilitent leur décision de voyager, de comprendre la façon dont les gens s'adaptent, négocient leurs contraintes potentielles et réelles et deviennent des voyageurs actifs. Les données de cette étude qualitative ont été obtenues à l'aide d'entrevues approfondies visant à donner une voix aux personnes ayant une déficience visuelle. Les résultats de l'analyse du contenu des données collectées mènent à ces conclusions importantes: La participation aux activités touristiques des personnes ayant une déficience visuelle résulte d'un processus continu et interactif qui s'appuie sur de multiples facteurs, avec des influences positives ou négatives sur le plan individuel, en tenant compte du contexte personnel, des conditions de dégradation et d'environnement social ayant un impact sur les différentes étapes du processus. Il est possible d'identifier les modèles de facteurs influençant le processus (contraintes et facilitateurs) ainsi que les stratégies de négociation en usage en vue de s'engager à la participation au tourisme. Pour surmonter les obstacles auxquels elles sont confrontées, les personnes handicapées utilisent les stratégies de négociation associées au processus d'organisation des voyages, ainsi que les différentes stratégies personnelles et interpersonnelles que cette étude a systématisées.

摘要

本研究主要侧重于理解残疾人士参与旅游活动的情况。具体而言，我们打算分析视觉障碍患者的经历，寻找约束因素以及促使他们决定出行的因素，试图了解人们如何适应、越过他们感知的和现实的约束，并成为活跃的旅客。定性研究采用深入访谈，旨在为视障人士发声。研究对收集的信息进行内容分析，分析结果可以得出以下主要结论：视障人士参与旅游活动的原因来自于持续的互动过程，这种过程由多种因素共同决定，当中有正面或负面的影响，这些发生在每个人的个人背景下，障碍情况和社会环境中，在过程的不同阶段的影响下。由此可以识别到影响过程的因素的模式（限制因素和促进因素）以及用于持续参与旅游的谈判策略。为了克服他们所面临的障碍，残疾人依靠与旅游组织过程相关的谈判策略，并依赖于在这项研究里系统化的不同的个人和人际策略。

Introduction

Although accessible tourism is evolving as an emergent field of academic research, the knowledge related to participation of people with visual disabilities in tourist activities is still incipient. In this sense, it is very important to understand the factors that shape their travel decisions and experiences. This can be an important contribution to the development of more accessible tourist products and destinations, giving tourism professionals a body of knowledge that allows them to respond with more appropriate strategies to address the needs of this group of people, and, thus, promote a more inclusive approach to tourism.

The results may also help to raise awareness among individuals with disabilities, their families and companions, as well as society in general, about the participation context of this group of people in tourism activities. In fact, there seems to be a need to develop new attitudes that encourage and support accessible tourism practices, identified as very beneficial for these individuals' personal development, social inclusion and overall well-being.

In this context, accessible tourism may present itself as a vehicle to promote individual and social well-being, not only for directly benefiting participants – a typically socially marginalized group – and their families, but also for society as a whole, by increasing social and family capital within the aforementioned groups, who in turn may become more empowered and active within the multifold dynamics of society at large. As has been highlighted by McCabe (2009), holidays provide people opportunities for positive change that influence their sense of well-being, their life experiences and horizons and could thus play an important role in promoting a healthy society; these opportunities should be even more relevant for individuals conditioned by an impairment. All these arguments justify initiatives of social tourism representing an investment in more inclusive, socially fair and healthy societies.

An enhanced understanding of the process that makes persons with disabilities more active and satisfied travelers is, in this context, crucial for developing effective public policies that yield more inclusive tourism opportunities, especially in the domain of social tourism programs. These could in fact help overcome many barriers and enhance this group's access to leisure and tourism, a sphere of life that is frequently taken for granted by most citizens in the developed world and considered fundamental to the quality of life, but which is still not equally accessible to all (Kastenholz, Eusébio, & Figueiredo, 2015).

Our approach is in line with the emerging paradigm of hopeful tourism scholarship, 'a new perspective which combines co-transformative learning and action to offer a distinctive approach to tourism knowledge production' (Pritchard, Morgan, & Ateljevic, 2011, p. 942). It advocates that knowledge, pedagogy, and action must be articulated in a way that promotes social justice (Richards, Pritchard, & Morgan, 2010), encouraging tourism research to focus on socially relevant themes that may result in more inclusive tourism practice.

Framed by these principles, the purpose of this paper is to understand the travel experiences of blind people, identifying the constraints and the facilitators that shape their travel participation as well as their decision-making process, forms of adaptation to and negotiation of travel constraints.

The paper begins with a discussion of the conceptual framework regarding the main themes discussed here: constraints and facilitators of participation in leisure and tourism faced by people with disabilities, and negotiation strategies used. It proceeds with explaining the methodology used in our empirical study, and then presents and discusses the main findings. The article ends with some conclusions and practical implications to policy-makers and tourism agents in order to develop conditions for more inclusive tourism initiatives, specifically considering the group of individuals affected by visual impairments.

Conceptual framework

Constraints to leisure and tourism

The research in the field of leisure constraints seeks to study 'factors that are assumed by researchers and perceived or experienced by individuals to limit the formation of leisure preferences and to inhibit or prohibit participation and enjoyment in leisure' (Jackson, 1997, p. 461). Several authors have been calling attention to the advantage of using the theoretical advances from the leisure constraints field to better understand the mechanisms of participation and decision making in tourism activities (Hinch & Jackson, 2000; Nyaupane, Morais, & Graefe, 2004).

The theoretical sophistication achieved by leisure constraints studies should, indeed, be a valid contribution to deepen the knowledge about the participation of people with disabilities in tourism. On the other hand, as several authors argue (Daniels, Drogin Rodgers, & Wiggins, 2005; Packer, Mckercher, & Yau, 2007; Yau, McKercher, & Packer, 2004), to better understand this market it is important not only to know the barriers and factors that prevent people with disabilities from traveling, but also to understand how they become active travelers, how they adapt to and negotiate the diverse constraints that emerge in that context.

According to the leisure research literature, there are three types of constraints: structural, interpersonal and intrapersonal, that have been consistently supported by empirical studies published in the field. Structural or environmental constraints intervene between the phases of forming preferences and effective participation (Crawford & Godbey, 1987). Constraining structural factors are associated with the broader, external context of the individual, which includes the lack of available time, financial constraints, transportation difficulties, lack of suitable infrastructures, among others.

According to Nyaupane and Andereck (2008), the main structural constraints identified in tourism literature may be classified into three sub-dimensions: time (Bialeschki & Henderson, 1988; Hung & Petrick, 2010; Nyaupane & Andereck, 2008; Pennington-Gray & Kerstetter, 2002; Williams & Fidgeon, 2000), financial resources (Bialeschki & Henderson, 1988; Fleischer & Pizam, 2002; Hung & Petrick, 2010; Nyaupane & Andereck, 2008; Pennington-Gray & Kerstetter, 2002; Williams & Fidgeon, 2000) and destination attributes (Blazey, 1987; Daniels et al., 2005; Fleischer & Pizam, 2002; Gilbert & Hudson, 2000; Hinch & Jackson, 2000; Nyaupane & Andereck, 2008; Pennington-Gray & Kerstetter, 2002; Williams & Fidgeon, 2000).

Interpersonal constraints can occur during interactions with an individual's social network, service providers or strangers, or because an individual lacks a partner with whom to engage in leisure activities (Crawford & Godbey, 1987). Studies have shown that many interpersonal factors can prevent people from participation in tourism activities. The most often cited include no companion (Bialeschki & Henderson, 1988; Daniels et al., 2005; Gilbert & Hudson, 2000; Hinch & Jackson, 2000; Hung & Petrick, 2010; Nimrod, 2008; Nyaupane & Andereck, 2008; Pennington-Gray & Kerstetter, 2002) and the influence of family and friends (Gilbert & Hudson, 2000; Nyaupane & Andereck, 2008; Pennington-Gray & Kerstetter, 2002).

Intrapersonal constraints are related to a person's psychological state, physical functioning or cognitive abilities (Crawford & Godbey, 1987; Smith, 1987), and include factors such as stress, anxiety, lack of knowledge, health related problems and social ineffectiveness. The most common intrapersonal constraints found in the tourism literature include health condition (Bialeschki & Henderson, 1988; Hung & Petrick, 2010; Nimrod, 2008; Nyaupane & Andereck, 2008), age (Bialeschki & Henderson, 1988; Fleischer & Pizam, 2002; Nyaupane & Andereck, 2008), personal fears (Gilbert & Hudson, 2000; Hung & Petrick, 2010; Nyaupane & Andereck, 2008; Nyaupane et al., 2004; Williams & Fidgeon, 2000), personal skills (Daniels et al., 2005; Fleischer & Pizam, 2002; Pennington-Gray & Kerstetter, 2002) and lack of interest (Nyaupane & Andereck, 2008).

In addition to the above mentioned constraints, people with disabilities are affected by many other factors that can influence their tourism decisions and experiences. In this paper, we focus only on those factors that have an increased importance in the context of the travel experiences of people with disabilities, leaving aside all others that are common to the population at large.

Although only the research conducted by Daniels et al. (2005) and by Small, Darcy, and Packer (2012) use the leisure constraints construct when analyzing tourism participation of persons with disabilities, the literature review allowed the identification of a set of constraints that seem appropriate in this context. However, most of the studies on this subject focus mainly on the population with physical disabilities, while the research on people with visual disabilities is still very scarce.

As far as structural constraints faced by people with visual impairment are concerned, the main issues identified in the published research are associated with: a lack of reliable information (Mesquita & Carneiro, 2016; Miller & Kirk, 2002; Small et al., 2012); rules and regulations (Small et al., 2012; Smith, 1987); lack of support services and of tourism providers' knowledge (Baker, Stephens, & Hill, 2002; Miller & Kirk, 2002; Richards et al., 2010; Small et al., 2012); lack of information; physical constraints encountered at natural and cultural attractions (Mesquita & Carneiro, 2016; Pearn, 2011; Richards et al., 2010), at restaurants (Small et al., 2012) or accommodation units (Baker et al., 2002; Small et al., 2012).

Interpersonal constraints result from interactions with an individual's social network, service providers or strangers. The main interpersonal factors affecting tourism participation of people with disabilities include: lack of encouragement from the family (Packer et al., 2007; Yau et al., 2004); dependence on others (Blichfeldt & Nicolaisen, 2011; Packer et al., 2007; Smith, 1987); negative attitudes from tourism providers (Smith, 1987; Yau et al., 2004) and from other tourists (Smith, 1987). Although their impact and intensity may vary according to the type of impairments, the interpersonal constraints identified in the scientific literature appear to be transversal to different disability groups. In fact, the few studies which analyze the travel experiences of people with visual impairment reinforce the significance of negative social attitudes, which often result from a lack of awareness and ignorance about disabled people's needs (Daruwalla & Darcy, 2005; Small et al., 2012). These stereotyped perspectives are also based on the assumption that people with disabilities are a homogeneous group with similar needs and characteristics (Daruwalla & Darcy, 2005; Richards et al., 2010).

In respect to intrapersonal constraints, the literature emphasizes the following factors: personality, motivations, emotions, personal fears, self-esteem, individual beliefs, perception of results from participation and previous tourist experiences. These factors seem to influence people with disabilities more deeply, resulting in feelings of vulnerability, anxiety and stress (Daniels et al., 2005; Darcy & Dickson, 2009; Packer et al., 2007; Richards et al., 2010) that also affect satisfaction with the tourism experiences and may prevent the desire to travel in the future.

Facilitators to leisure and tourism

According to Raymore (2002), it is most important to understand the nature of participation in leisure activities to comprehend not only the factors that constrain it but also the facilitating factors and how they combine to foster participation, or non-participation, as well as the resulting experiences. In this sense, the systematic approach distinguishing and analyzing the nature and interaction between leisure facilitators and constraints may significantly contribute to a better understanding of the dynamics of participation or non-participation in tourism.

In line with Raymore (2002), facilitators and constraints are not alternative explanations for justifying participation but rather complementary approaches to understanding involvement in leisure. On the other hand, the relevance of the various dimensions of facilitators and constraints varies from activity to activity and from individual to individual; in this sense, people face different constellations of facilitators and constraints which, when combined, will determine the decision to travel.

Similar to constraints, facilitators may also belong to one of three dimensions: interpersonal, intrapersonal and structural, promoting the formation of preferences and encouraging participation. Specifically, regarding the factors that influence the decision to travel for people with disabilities, research has focused mainly on restrictive factors. Nevertheless, it is possible to also highlight a set of facilitating factors, although they have not been, to date, studied in the light of the above mentioned leisure participation framework.

In terms of interpersonal facilitators, it is possible to identify the following factors: stimulus and support from an individual's social network, interaction with tourism sector professionals, travel companions, and positive attitudes of others. According to Packer et al. (2007), the encouragement and support of family and other people in their social network play a decisive role, supporting the decision-making process and helping to overcome some structural barriers. Additionally, well-trained, attentive and helpful staff can be a facilitating factor in the tourist experience of people with disabilities, as opposed to negative attitudes. In this regard, Packer et al. (2007) emphasize the role of positive and attentive attitudes of professionals as a significant facilitator which can, to a large extent, overcome some of the structural barriers affecting participation of this group of people in tourism activities. Travel companions are one of the most cited facilitators in published research, assuming a significant role in the travel decision process (Devile, Kastenholz, & Santiago, 2012; Packer et al., 2007; Yau et al., 2004).

As for intrapersonal facilitators, several studies emphasize the role of personal characteristics for the way people with disabilities overcome the adversities associated with a disability condition. Acceptance of a disability and a person's self-confidence thus emerge as facilitating factors that enable them to cope with difficulties, whether in their daily life or when they intend to travel (Daniels et al., 2005; Devile et al., 2012; Packer et al., 2007). The study conducted by Blichfeldt and Nicolaisen (2011) suggests that there is a strong inter-dependence between an active life and the participation of people with disabilities in tourism activities. On the other hand, a person who travels regularly becomes a more experienced tourist, which fosters confidence in their tourism consumption decisions (Blichfeldt & Nicolaisen, 2011; Yau et al., 2004). As a result, previous tourist experiences and the accumulated knowledge about tourism allow the individual to deal with obstacles in a more positive and knowledgeable way, which increases their feelings of safety and stimulates their motivation and desire to travel.

Structural facilitators encompass a set of factors related to the availability of accessible tourist services, which can encourage people with disabilities to participate in tourism. Research in this area focuses mainly on issues associated with accommodation. Neverthe-less, it is possible to deduce from the identified constraints that these are not restricted to accommodation but to the overall accessibility of tourism services – attractions, transpor-tation and restaurants, among others.

The existence of proper accommodation with adequate accessibility conditions emerges as a critical factor in the tourism decision-making process for people with disabilities (Daniels et al., 2005), being crucial to ensure a comfortable stay and to allow their inde-pendence from other people. Another important facilitating factor is the availability of reliable and accurate information in order to meet tourists' differential requirements and allow them to make informed travel decisions, thus avoiding unpleasant surprises and unnecessary risks (Bieger & Laesser, 2001; Cavinato & Cuckovich, 1992; Daniels et al., 2005; Eichhorn, Miller, Michopoulou, & Buhalis, 2008; Packer et al., 2007).

Negotiation strategies

The concept of negotiation was introduced by Jackson, Crawford, and Godbey (1993) in leisure studies, based on the underlying idea that people develop efforts to overcome con-straints, through cognitive or behavioral strategies that further leisure participation. That is, people often adopt innovative strategies to mitigate the effects of constraints, either by modifying leisure habits or by changing other aspects of their lives (Henderson, Bedini, Hecht, & Schuler, 1995; Jackson & Rucks, 1995). According to Jackson (2000), the strength of leisure motivations, and the perceived importance of anticipated benefits, encourages people to try and succeed in this negotiation.

The idea of negotiation of constraints is consistent with the socio-cognitive perspective, whereby people respond to conditions that prevent them from achieving their goals rather than accepting them passively (Loucks-Atkinson & Mannell, 2007). According to Jackson and Rucks (1995), negotiation may, for example, involve cognitive strategies that alter the perceived value of leisure activity, or behavioral strategies such as adjustment of schedules, reassessment of priorities in terms of money, time, energy, and information strategies. So, the negotiation process is different from constraints acceptance, implying efforts to change a situation, which will allow a compromise to be found or a problem to be solved, resulting in more positive meanings of the constraint and recognition of new opportunities, within a wider social context (Hutchinson & Kleiber, 2005). On the contrary, if a person chooses to avoid an activity, for instance due to a missing companion, this would lead to accommo-dating to the situation and not to a constraint negotiation.

Table 1. Summary of negotiation strategies used by people with disabilities identified in the scientific literature.

Constraints		Negotiation Strategies	Authors
Structural	Accommodation	Travel Preplanning	Daniels et al. (2005), Blichfeldt and Nicolaisen (2011)
	Transportation	Information search on service	
	Tourist attractions	providers	Daniels et al. (2005), Blichfeldt and Nicolaisen (2011), Yau et al. (2004)
		Rights argumentation	Richards et al. (2010)
		Help from strangers	Daniels et al. (2005)
	Lack of information	Use of multiple strategies to confirm the information	Packer et al. (2007), Yau et al. (2004)
Intrapersonal	Feelings of vulnerability	Mental fortitude	Daniels et al. (2005), Richards et al. (2010)
		Self-determination	
		Self-reliance	
		Concentration and mental memory	
Interpersonal	Dependence on family and friends	Help from strangers	Daniels et al. (2005)
		Travel companion	
	Negative social attitudes	Traveling with other people with disabilities	
		Traveling with friends and family	Yau et al. (2004)
		Traveling with other people with disabilities	

Negotiation strategies used by people with disabilities in the tourism context constitute an underexplored research area with scarce information on how constraints are related to participation or how different constraints are balanced with negotiation efforts. From our literature review, Daniels et al.'s (2005) qualitative study stands out as the only approach focusing on constraints to tourism participation, seeking to understand how individuals with disabilities react and adapt to different constraints. The authors identified six intrapersonal, six interpersonal and eight structural constraints, suggesting the need to use negotiation strategies to overcome each of them. Despite the important findings of this research, some of the negotiation strategies identified are facilitators rather than negotiation strategies, insofar as they refer to existing conditions both in the individual's personal sphere and in their external environment. This is the case where, for example, positive attitudes of the staff are identified as a negotiation strategy, but seem rather classifiable as a facilitating factor, which, at best, may lead to the adoption of a negotiation strategy (for example, asking for help) to overcome a constraint.

Other studies have contributed to the understanding of different mechanisms used by people with disabilities to overcome barriers in their tourism activities, even though not adopting the negotiation construct explicitly. Table 1 summarizes the constraints and main negotiation strategies identified in the literature review in this field. Most of the studies focus on trip organization issues, emphasizing the importance of rigorous and detailed planning of a trip, and the selection of service providers.

Methods

The empirical work employed a qualitative methodology, which sought to give voice to people with visual disabilities. This research intends to contribute to a deeper understanding of the feelings and perceptions of tourists with disabilities. For this purpose, the most suitable technique of data collection was the in-depth interview. This is the right tool to create a space for dialogue as it favors a natural environment of communication and sharing of the interviewees' ideas and opinions. On the other hand, it also has the advantage of obtaining information on the meaning that the actors attribute to their practices and the events with which they are confronted, allowing a high degree of depth of the data collected. Conversations with the individuals in the target group reveal the reflections of these actors on their own tourism behavior, uncovering what it means to be a disabled tourist, as well as why and how they make their travel choices, within the individual context and lifestyle.

Several authors have drawn attention to the need to 'give voice to people with disabilities' (e.g. Blichfeldt & Nicolaisen, 2011; Kitchin, 2000; Richards et al., 2010) in order to achieve a more in-depth understanding of their feelings and perceptions during their tourist experiences, beyond physical accessibility conditions of the places they visited (Blichfeldt & Nicolaisen, 2011). According to Richards et al. (2010), researchers should seek to explore new ways of understanding accessible tourism by giving a central role to people with disabilities in the research process, and thus promote positive social change. This line of thought, framed by emancipatory disability research, is situated in the paradigm of so-called hopeful tourism research. Recognizing the close articulation between theory, action and practice, this paradigm assumes that ethical obligations are intrinsic to research, yielding knowledge that can be translated into socially inclusive and fair tourism practices (Ren, Pritchard, & Morgan, 2010; Richards et al., 2010; Sedgley, Pritchard, & Morgan, 2011).

In order to obtain a holistic understanding of the tourist experiences of people with disabilities, respondents should be allowed to explain how their disability influences their tourism decisions, in a relatively unconstrained manner. For that reason, a semi-structured questionnaire was developed which explored several main concepts through brief questions (mainly regarding the travel context and conditioning factors for travel decisions), and which conferred considerable freedom of speech to the interviewees who were encouraged to explore the suggested topics referring to their own complex and idiosyncratic experiences (Quivy & Campenhoudt, 1992).

The identification of the interviewees was initially based on our personal contacts, resulting from our previous work in this field, linking us to the interviewees and to other social actors who mediated additional contacts. The selection of participants yielded diversity in terms of travel patterns, type and degree of disability, age, income and gender. These selection criteria allowed us to identify different factors and analyze patterns according to different situations and lifestyles. Individuals were assured anonymous data treatment and presentation, were previously briefed on the objectives of the study and were totally free to respond – or not, if they were uncomfortable about a topic – in any manner they considered appropriate. In total twelve adults of varying ages (between 32 and 80 years) with visual impairment agreed to share their views and travel experiences. Most respondents had acquired blindness as a result of progressive congenital disease (macular degeneration), two individuals were blind from birth, and in one case blindness resulted from a war accident. Of the twelve blind people interviewed, eight used guide dogs. As for their family situation, eight respondents were married and the remaining four were single or divorced and lived alone. All the respondents, except for two retired individuals, were professionally active.

We may therefore consider that, in a first phase, a purposeful sampling technique was applied, oriented by the objective of covering a diversity of situations in the studied universe. At a later stage, the 'snowball' technique was used to identify other potential participants, that is, as we were doing the interviews, we asked our interviewees to provide us with contacts of people who could collaborate in this study. Respondents were very receptive and available to collaborate, which f allowed an atmosphere of empathy and trust creating a positive environment for conducting interviews. It became clear that the actors considered their opinion important to eventually introduce changes in a domain of their personal relevance (Quivy & Campenhoudt, 1992).

The interviews were tape-recorded with the participants' permission and fully transcribed. This process, although very time consuming, allows for greater interaction and flexibility in conducting the interview and provides a more accurate record of the information given by the interviewees (Jennings, 2005; Silverman, 2000). Transcription of 15.5 hours of interviewing resulted in a total of 398 pages of transcribed text. The transcripts were emailed to each participant to obtain feedback, generally aided by an intermediary. Based on their responses, a few changes were made, including the elimination of some details to ensure confidentiality.

The data was treated via content analysis identifying major themes and categories in the discourses. This process involves three phases: (a) pre-analysis (organization of the material and definition of procedures); (b) exploration (identification of categories and intersections, with narrow amplitudes and connections); (c) treatment of results (the data is interpreted and gains meaning). A software supporting qualitative data analysis, namely WebQda, was used to create categories, codify, control, filter, search and query the data.

The process of information codification, i.e. the production of a system of categories, was based on a mixed method, resulting both from the conceptual framework and the empirical material constituted by the discourse of the actors, in a dynamic, hybrid deductive and inductive process of confrontation between theory and evidence flowing from the discursive material (Silverman, 2000). As new themes emerged, new fields were added or dismembered in order to capture the richness of the discursive material relevant to the present research purpose.

Findings and discussion

Major constraints and negotiation strategies used

In our study, structural constraints were the most often mentioned, with nine structural constraints identified. The meanings of each one as well as the negotiation strategies used to overcome them are summarized in Table 2.

Lack of knowledge of tourism providers emerged as an important constraint across all narratives, being reflected in different ways as distinctly impacting on travel experience and satisfaction. Some interviewees recognized that, in most cases, this constraint reveals a lack of training and ignorance on how to deal with blind guests and help them, as described in the following quote:

> They have goodwill, but much ignorance. ... They do not know how to help ... sometimes they have sensitivity, they are willing to welcome the blind passenger, but then they do not know how to help! Usually people tend to push the blind person in front of them, but it is not the blind person who has to go ahead; the blind person needs to go behind the guide ... .They put the hand on my back, they push me. Now, if I have to

Table 2. Structural constraints and negotiation strategies used.

Structural constraints	Meaning	Negotiation Strategies used
Service providers	Lack of attention to customers with disabilities Lack of training of professionals, considering people with disabilities as a homogeneous group.	Mental fortitude
Tourist attractions	Lack of equipment and services to improve sensory or communication accessibility: audio guides, tactile experiences, Braille publications and labeling.	Selection of service providers Information sources Notice in advance Travel companion Adaptation ability
Information availability	Lack of information on accessibility Lack of accessible websites Lack of information in Braille	Information sources
Restaurants	Difficulty in buffet services Lack of menus in Braille Difficulty in handling the food served Resistance to guide dogs	Travel companion Help from strangers Rights argumentation
Rules & regulations	Barriers to the use of guide dog, especially in air transportation, either forcing the transport in the aircraft hold or requiring the use of a muzzle	Rights argumentation
Accommodation	Disposal of facilities and furniture in the bedrooms	Notice in advance Adaptation ability
Transportation	Difficulties in moving in the metro and train stations; Resistance to the use of guide dogs Difficulty of using automatic ticket machines	Rights argumentation Travel companion
Lack of support services	Lack of tourist guides	not identified
Public space	Lack of accessible routes Lack of uniform standards	not identified

fall, I am the one to fall first, so they do not know. They have a lot of goodwill; they have pity; they have this curiosity, this whole goodwill, but they have no training. They show a lot of ignorance. (A1)

The narratives also confirm that it is common to consider people with disabilities as a homogeneous group with similar needs, as recognized in previous studies (Daruwalla & Darcy, 2005; Richards et al., 2010). The negotiation strategy used to deal with this involves a positive attitude and mental fortitude when trying to face adversities in a proactive way, adapting to situations and not letting them negatively affect the tourist experience. *With my humility, with my simplicity. There, that's how I do it, and that's how you learn, you see? Because it's not with arrogance, it isn't with bad manners, no, it's not like that; it's really just simplicity (A8).*

The other structural constraints identified are mainly associated with the lack of services and equipment in the tourism offer, which affects people in different ways, depending on whether they travel alone or with a companion or other people with similar impairments, as is the case with eight participants. In these cases, the main negotiation strategy is rigorous travel planning, undertaking a huge travel information search to verify the service conditions necessary to meet their needs. This includes anticipating and preventing problems in advance, especially regarding accommodation and transportation, as one of our interviewees notes:

When I make a trip, I say, 'attention, I'm visually impaired; I'm going to need help and I'm accompanied by a guide dog' ... .Even if I have problems, I do not want to be stressed when I arrive at the hotel, so if I have problems, I want to be prepared with all the relevant documentation, and I want people to be informed so that there is no confusion. (A10)

Doing this can also be a test to verify the attitude and sensibility of service providers. For instance, one traveler reports, *First I contact, I say 'I am a person with visual impairment, do you think there are conditions for . ..' and I see if they have that sensitivity* (A3).

It is also important to note that despite the legislation on the use of guide dogs, there are still some barriers, especially in air transportation. In these cases, the main negotiation resource is for the traveler to defend and assert their rights in order to resolve the situation they face and also to draw attention to discriminatory practices.

Sometimes, when a situation cannot be resolved, respondents admit to filing a complaint, calling the police or even resorting to the courts; as one respondent, who was prevented from entering a restaurant with their guide dog, stated: *Simply, they did not want the dog to come in. I explained, I don't know if they realized or not. The truth is I couldn't convince them; I had to call the police* (A9). Another respondent, whose dog a taxi driver refused to transport, had the same attitude,

> And then I said that I'll call the police and he said 'then call, I'll stop'. But he didn't. I leaned against the door and he drove away. Somebody took the license plate number and I took him to court. (A5)

The most emphasized interpersonal constraints, presented in Table 3, are: negative social attitudes, dependence on family and friends, attitudes of staff, erroneous perception of disability, fear of disturbing others and constraints resulting from family context.

The prevalence of negative attitudes of others is often associated with social preconceptions and stereotypes, resulting in large part from the lack of information and knowledge about the reality of people living with disabilities (Daruwalla & Darcy, 2005). Virtually all study participants indicated that they face various kinds of negative attitudes, both from social interaction with other tourists as well as from tourism staff. The discourses revealed a certain discomfort regarding the way other people interact with them, either by adopting a paternalistic attitude, or reflecting some social inhibition by opting to ignore them,

Table 3. Interpersonal constraints and negotiation strategies used.

Interpersonal constraints	Meaning	Negotiation Strategies used
Negative social attitudes	Paternalistic Actions	Mental fortitude
	Lack of spontaneity in social interaction	Positive attitude
	Avoid direct social contact	Adaptation ability
	Attitude toward the guide dog	Rights argumentation
Dependence on family and friends	Difficulties in moving at unfamiliar surroundings	Help from strangers
	Need for someone to provide personalized information	Selection of services providers Travel companion
Negative attitudes from tourism staff	Social inhibition	Mental fortitude
	Unhelpful attitude	Positive attitude Notice in advance
Erroneous perception of disability	Consider people with disabilities as a homogeneous group	Mental fortitude
	Social stereotypes about blind person	Positive attitude
Fear of disturbing others	Psychological discomfort	Help from strangers
	Feeling of vulnerability	selection of services providers
	Increased burden for caregivers at holiday time	Travel companion
Family context	Fears expressed by the family	Mental fortitude
		Positive attitude

addressing the companion instead; this behavior is also found in other studies (e.g. Daruwalla & Darcy, 2005). *When we are with another person who does not have any type of impairment, generally, no matter how many times we ask the question, the interlocutor is always the other person. This annoys me a lot (A12).*

An erroneous perception of disability refers to the assumption that people with disabilities are defenseless, depressed and victims of a personal tragedy (Daruwalla & Darcy, 2005), highlighting the existence of frequent social stereotypes about blind persons. Sometimes curiosity is aroused, especially regarding their interest in traveling, with one interviewee commenting:

> Because they always see [the travel motives of] a blind person like: 'oh, he goes out of curiosity; he goes because he has nothing else to do, he goes because he has been dragged there, because he can go here as much as to any other place'. This is … not believing in the maturity and intellectual component that a blind person may have. Because she said it with all the spontaneity, in fact she didn't say that to offend. (A1)

To avoid the negative impact of these episodes on the tourist experience, our interviewees use strength of mind (or mental fortitude) and self-esteem to overcome adversity, as identified by Daniels et al. (2005), as an emotional negotiation strategy. This attitude of self-confidence is also reflected in the way they interact with others, how they try to understand what underlies certain negative attitudes, as emerges in the following account,

> I understand that it is an unusual situation for them. I think it makes them think, something, probably it's not necessarily bad, you see? I think this is a confrontation with human fragility and eventually they think, poor him, and the luck he has, and that's it. I'm not poor, I feel fulfilled and therefore I do not let myself be affected by it. (A12)

The dependence on family and friends is associated with a lack of autonomy (intrapersonal constrain) and a need for assurance and for practical help during a trip (Yau et al., 2004), mostly expressed by people who do not use a guide dog.

> I cannot go abroad alone. I have to be accompanied by someone who is willing to help making trips and finding places, service locations, hotels, moving in hotels, moving on walks. These are completely unknown spaces and therefore a blind person does not have autonomy, if they are alone; so they have to go with someone who is available, not only to do this work as a guide, but also with another characteristic that I think is fundamental, and here I have been lucky enough to meet people . . some of whom have somehow lent me their eyes, described landscapes to me, perceived my tastes, what I like. (A1)

As illustrated in this narrative, the travel companion selection is one of the strategies used to overcome this constraint. Although those who use guide dogs recognize the importance of having a non-disabled travel companion, they do not stop traveling when they have none.

Some of the interviewees express fear of disturbing others, resulting in psychological discomfort and feelings of vulnerability. *Sometimes I feel disconsolate. I say, I could go alone; I'm always bothering those persons, it's a bit like that, it costs me! (A8).* In order to avoid being an increased burden for caregivers at holiday times, the actors try to get help from strangers or select the service providers that best meet their needs.

Familial context emerged sometimes as an interpersonal constraint in so far as the lack of encouragement, or the fear expressed by a family about the risks of traveling may cause some apprehension and discourage travel. According to former research (Packer et al., 2007; Smith, 1987; Yau et al., 2004), a family's lack of incentive to travel can result from an overprotective role in relation to the person with disability. However, this family context can also function as an important facilitator as we will discuss later. This is most strongly felt when it comes to a newly acquired disability or to early tourist experiences: *The family said 'but you're going to travel? But how do you go alone? You cannot go, you have to go with your wife!' That's a burden, it's a very big weight (A4).*

As suggested by Loucks-Atkinson and Mannell (2007), our analysis also seems to show that people with high negotiation effectiveness, that is, with confidence in their ability to use negotiation strategies and resources, are more motivated to participate in travel, overcoming barriers through negotiation. In this sense, the use of the self-efficacy construct in negotiation helps to explain how those more motivated to participate will have a lower constraints perception, stimulating the negotiation efforts which indirectly positively influence participation (White, 2008).

In respect to intrapersonal constraints, our research identifies the following categories: perception of physical disabilities, risk perception, discomfort, stress and lack of autonomy (Table 4). While admitting that disability does not prevent them from traveling, most interviewees recognize that they would travel much more and differently if they did not have the limitations imposed on them by disability.

> I would really love to do trekking in the Himalayas; it's something I would very much like to do. But I cannot do that. Neither can I do it with my dog; I do not know if my dog could adapt to the altitude. I cannot do it. But this kind of travel I would very much like to do. (A11)

> Therefore, visual impairment, yes, conditions me, because I have no doubt that if I could see, I would naturally have a driving license and would move much more autonomously and much more to where I liked to go and how I liked to do it. So, I am always conditioned, as you see, I always have to be touching ground, and organizing, circumscribing contexts and organizing spaces and evaluating. For me it's more complicated. (A3)

Besides the inability to enjoy visual experiences from journeys, the main problem identified by our respondents is related to orientation in unfamiliar spaces. As noted by Small et al. (2012), an unfamiliar travel environment creates feelings of anxiety and insecurity due to a lack of sense of control. The risk perception is then a cause of discomfort and stress, mainly resulting from anticipating problems and – out of necessity – having to concentrate when navigating in new surroundings.

Table 4. Intrapersonal constraints and negotiation strategies used.

Intrapersonal Constraints	Meaning	Negotiation Strategies used
Perception of disability	Limitation on travel choices Inability to enjoy visual experiences Impossibility to drive Orientation difficulty at unknown spaces	Travel companion Help from strangers Mental fortitude Positive attitude
Risk Perception	Feelings of insecurity at unfamiliar environments	Travel companion Sources of Information/ preplanning travel
Discomfort and stress	Constant necessity of concentration and adaptation in unfamiliar surroundings Feelings of anxiety and concern resulting from anticipation of problems in unfamiliar environments	Selection of places to visit Selection of service providers Travel companion
Lack of autonomy	Lack of freedom and independence Orientation problems in unfamiliar environments Impossibility of using mental schemes to guidance	Selection of services providers Travel companion Ability to adapt

The travel companion selection is once again one of most common strategies used to overcome this constraint. The participants who travel alone or with other people with visual impairment negotiate these situations in different ways: preplanning travel (information search), selecting the places to visit and the service providers to use.

> Before traveling, if I go with a group of blind people, I will first do research on everything that is around, which is for later, when we arrive, [I try to] get to know everything there is to do and how we can go. … I'm a person who always asks where and how to get there. … I'll have to choose very well the hotel I want: first, if I want to go walking, it has to be as central as possible; then if the hotel allows me to travel daily in their transports, because sometimes they have certain routes but they have no other type of transport, and also public transportation is not accessible to this hotel other than the taxi, so I will always think of this when choosing a hotel. (A10)

> I really prefer quieter [destinations]. We always choose places that are as calm as possible, because it makes it easier for us to orient ourselves, then after a day or two, we'll also get to know the space better and we'll have more autonomy. (A9)

Major facilitators to travel

The decision to travel also results from the interaction of the facilitating factors, which, in contrast to constraints, promote and encourage participation in tourism. Table 5 presents the categories and themes identified in our study.

Table 5. Facilitators to travel.

Intrapersonal facilitators	Meaning
Acceptance of disability	Proactive attitude
	Self-reliance
Determination	Perseverance to achieve their goals
	Ability to overcome adversities during travel
Previous travel experiences	Influence of travel experiences during childhood
Personality	Pleasure of learning and discovering new places
	Spirit of adventure
Curiosity	Pleasure of knowing new cultures
	Openness to new experiences

Interpersonal facilitators	Meaning
Interactions with tourism staff	Assertiveness and helpful attitude
	Spontaneity in social interaction
Travel companion	Assistance during travel meandering
	Suitable personal traits towards people's needs
	Desire to visit family and friends living abroad
Encouragement and support from the social network	Incentive for autonomy and encouragement to travel
	Support on travel planning & organization
Positive attitudes of others	Helpful attitude from strangers

Structural facilitators	Meaning
Touristic attractions	Equipment to improve sensory accessibility (audio guides, models of monuments and replicas of works of art)
	Availability of personalized services
	Existence of Braille information
	Possibility of tactile experiences
Transportation services	Support services at airports and during flights
	Exemption from payment by the accompanying person.
	Proper functioning of trains services
Accommodation	Differentiated floors and elevators with sound system
	Information in Braille in the elevators and in the rooms doors

Intrapersonal facilitating factors, such as acceptance of disability, determination, previous travel experiences and personality, influence, to a large extent, the way respondents perceive constraints and how they choose to cope with them. The way people represent themselves has a significant influence on their tourism practices, adopting proactive and assertive attitudes that allow them to deal with and overcome the constraints they face.

In this context, the acceptance of disability, as Yau et al. (2004) point out, is of crucial importance. This often involves a lengthy process of personal learning, self-discovery, self-confidence and trust based on the recognition of difficulties and, above all, on the belief in their capacities and not in being a burden to others. The following testimony is a good example:

> First notes of autonomy can, in the course of a person who is blind, be the autonomy of eating, dressing, going to a hotel. A blind person cannot be a burden to himself or to the family, so it is necessary to invest … It was not easy at the initial stage, it was a long process … I have to work, I have to be a father, I have to be a husband, … and I also have to be myself. (A4)

Most of the respondents assume, without complexes, their differences, continuously seeking to develop a greater autonomy, and trying to change the disability prejudices and social stereotypes.

> And that is why I tell you that is important for us to show that we are able, and that it isn't as difficult as people think, but it is also necessary to have a little calm and to help people understand, isn't it? (A8)

> I think there is also an aspect whereupon I feel some responsibility: if we want society to better understand blind people we have to have an active attitude to contribute to this change. (A1)

The person's determination, curiosity and the meaning attributed to the pleasure of discovering new cultures positively influence people to travel and to overcome the many obstacles they still face.

> I really enjoyed this trip, we went to Switzerland, Germany, Austria, then returned to Italy. It was a trip that I really enjoyed, and then it awakened me a lot more for the trips. I found, that in fact, if we are at home we hear a kind of thing, but [what we find] is very different … . I like human contact, to know how people live, this for me is, in fact, enticing. … But I did not content myself to be told … no, I want to be shown everything that I keep in my mind, the detail of the perspective. (A8)

> I want to know things, I want to see what it's like, I want to go and feel it. (A10)

In the domain of interpersonal facilitators, the findings point out several factors – interaction with professionals in the tourism sector, travel companions, encouragement and support from the social network, and positive attitudes from others – that exert a positive influence on participation in tourism. Although some of these factors have also been identified as constraints, some respondents clearly acknowledge, for instance, the assertiveness and helpful attitude of tourism staff and other people as a relevant facilitator.

> For example, when we request assistance for blind people at the airport, it turns out that they know how to guide a blind person, what they can do, what they cannot do; therefore, I think they have some training there. (A11)

The role of the family in encouraging the respondent's autonomy and sense of discovery is also a facilitator in the decision to travel. *I try to face things naturally, a lot because since I was a little kid, my parents have been trying to encourage me to do this autonomously (A12).*

With regard to structural facilitators, our study highlighted mainly accessibility in touristic attractions, such as tactile experiences, audio guides, personalized services, access to information in Braille. *Some interesting things have been done, such as providing differentiated floors for the visually impaired, Braille information, and having audio devices to support the exhibition (A11).*

> There are already many lifts with sound systems, a Braille system on each floor, both in the elevators, inside and outside, and even in the room numbering itself. (A10)

Also accessible transportation services and inclusive accommodation were highlighted, with differentiated floors and elevators with sound systems, or with information in Braille. Transportation was also focused, mainly regarding support services at airports and during flights, exemption from payment for the accompanying person and proper functioning of train services.

Conclusions and implications

The findings of this study underline the dynamic and interactive nature of tourism participation by people with visual impairments who face different, interacting factors, which arise from their individual context, the social environment and the tourism context. It was possible to systematize the factors that constrain participation in tourism, revealing the prevalence of structural factors, mainly related to a lack of proper tourism services and awareness from the tourism industry regarding these people's needs. In fact, the predominant constraint reported was a lack of knowledge of tourism providers, having a strong impact on travel experience and satisfaction. Additionally, negative attitudes of others seem to be the interpersonal constraint that was most felt in the tourist experiences of people with visual impairments, apparently associated with prejudice and ignorance about the disability and its contours (Daruwalla & Darcy, 2005).

On the other hand, travel skills acquired by experience allowed for more confidence in tourism participation (Blichfeldt & Nicolaisen, 2011; Yau et al., 2004). Our research suggests that most travel constraints are perceived as a learning opportunity, thus enabling people, after successful coping, to make more informed decisions in the future and to avoid potentially risky situations.

Our research also permitted an understanding of the facilitating factors that influence travel decisions and satisfaction. Previous research in this field has clearly given more attention to the constraints factors. Our study shows that there are a set of facilitating intrapersonal, interpersonal and structural factors that are decisive in encouraging travel and should therefore be considered when designing accessible tourism strategies. Besides the influences of the personal sphere (personality, previous travel experience, determination, curiosity, self-confidence, acceptance of disability), our findings identified as relevant interpersonal factors: encouragement and support from the social network, travel companion, tourism staff awareness, as well as structural factors associated with accessible information, proper services and attractions that meet the needs of persons with disability.

People with disabilities use different mechanisms to overcome constraints that may occur at any phase of the tourism experience (Daniels et al., 2005). Our study revealed a set of negotiation strategies used by persons with disability, not previously considered in the literature, which were further systematized into personal strategies, travel organization strategies and interpersonal strategies. These strategies need to be understood, since they permit a more sophisticated understanding and targeting of this group, promoting more confident travel decision making and more fulfilling experiences.

This study additionally corroborated the lack of the tourism industry's knowledge of the quality of accessible tourism experiences for people with different disabilities (Figueiredo, Eusébio, & Kastenholz, 2012), drawing attention to the importance of the responsibility shared by the actors in the tourism system to promote and implement inclusive tourism. This requires a thorough knowledge about the different dimensions of the disability, also considering diversity within this group, and suggests the need for general training and awareness in this field at different levels. Unlike most studies on accessible tourism, we focus here on a specific group, namely people with visual impairment, reporting specific needs, constraints, facilitators and negotiation strategies.

In this sense, we hope to have added relevant insights into a complex specific reality, the understanding of which is crucial to improve tourism suppliers' capacity of attending to this group's needs and addressing their particular conditioning context, so as to help them overcome barriers and effectively become more active travelers. As several studies point out, this market may indeed become a particularly valuable target group due to its high levels of loyalty toward those who may best address its conditions.

Additionally, our results should be important to policy-makers concerned with the development of a more inclusive society (Kastenholz et al., 2015), who may identify relevant issues so as to ensure public policies and effective action allow quality tourism for all. Indeed, as we have seen, part of the difficulty involved in the full participation of people with visual impairment in tourist activities is due to the lack of compliance with legislation. However, more than regulatory mechanisms, there seems to be a need to mobilize collective responsibility, in tourism and other sectors, to promote a more just and inclusive society.

The implications for tourism policy drawn from our results include the need to promote disability awareness programs. For tourism providers, our results suggest the concrete need for more and better training of tourism staff regarding accessible tourism requirements and specific disability conditions. From a different perspective, it is also essential that the educational sector, at all levels, includes disability subjects on its curricula, creating awareness and sensitivity regarding each person's role in creating a more inclusive society. More specifically, it could be very useful to foster interaction with people with disability. As documented by Daruwalla and Darcy (2005), the contact could be very effective and powerful regarding an attitude change and, in this sense, tourism could be an enabler to attitudinal and behavioral change. There seems to be much room for developing social tourism programs targeting diverse groups of people with disabilities, with differentiated initiatives, staff training and inclusion of experiences that were identified as most appealing and gratifying, while also addressing the needs of those accompanying persons with visual impairment.

In the end, we hope that our work can contribute to more inclusive tourism practices and encourage a greater involvement of tourism researchers on disability issues, promoting social inclusion, human dignity and human rights of people with disabilities, thereby recognizing their differences and the need to integrate human diversity into the tourism system.

Disclosure statement

No potential conflict of interest was reported by the authors.

ORCID

Eugenia Devile ⓘ http://orcid.org/0000-0003-1643-5394

References

Baker, S. M., Stephens, D. L., & Hill, R. P. (2002). How can retailers enhance accessibility: Giving consumers with visual impairments a voice in the marketplace. *Journal of Retailing and Consumer Services, 9*, 227–239.

Bialeschki, M. D., & Henderson, K. A. (1988). Constraints to trail use. *Journal of Park and Recreation Administration, 6*, 20–28.

Bieger, T., & Laesser, C. (2001). Segmenting travel on the sourcing of information. In G. I. Crouch, J. R. B. Ritchie, & A. G. Woodside (Eds.), *Consumer psychology of tourism, hospitality and leisure* (pp. 153–167). Wallingford: Cabi Publishing.

Blazey, M. (1987). The differences between participants and non-participants in a senior travel program. *Journal of Travel Research, 26*(1), 7–12.

Blichfeldt, B. S., & Nicolaisen, J. (2011). Disabled travel: Not easy, but doable. *Current Issues in Tourism, 14*(1), 79–102.

Cavinato, J. L., & Cuckovich, M. L. (1992). Transportation and tourism for the disabled: An assessment. *Transportation Journal, 31*(3), 46–53.

Crawford, D., & Godbey, G. (1987). Reconceptualizing barriers to family leisure. *Leisure Sciences, 9*, 119–127.

Daniels, M. J., Drogin Rodgers, E. B., & Wiggins, B. P. (2005). 'Travel tales': An interpretive analysis of constraints and negotiations to pleasure travel as experienced by persons with physical disabilities. *Tourism Management, 26*, 919–930.

Darcy, S., & Dickson, T. (2009). A whole-of-life approach to tourism: The case for accessible tourism experiences. *Journal of Hospitality and Tourism Management, 16*, 32–44.

Daruwalla, P., & Darcy, S. (2005). Personal and societal attitudes to disability. *Annals of Tourism Research, 32*(3), 549–570.

Devile, E., Kastenholz, E., & Santiago, R. (2012). Inibidores, facilitadores e estratégias de negociação associadas às práticas turísticas das pessoas com incapacidade. *Revista Turismo & Desenvolvimento, 17/18*(3), 1417–1430.

Eichhorn, V., Miller, G., Michopoulou, E., & Buhalis, D. (2008). Enabling access to tourism through information schemes? *Annals of Tourism Research, 35*(1), 189–210.

Figueiredo, E., Eusébio, C., & Kastenholz, E. (2012). How diverse are tourists with disabilities? A pilot study on accessible leisure tourism experiences in Portugal. *International Journal of Tourism Research, 14*, 531–550.

Fleischer, A., & Pizam, A. (2002). Tourism constraints among Israeli seniors. *Annals of Tourism Research, 29*(1), 106–123.

Gilbert, D., & Hudson, S. (2000). Tourism demand constraints - a skiing participation. *Annals of Tourism Research, 27*(4), 906–925.

Henderson, K. A., Bedini, L., Hecht, L., & Schuler, R. (1995). Women with physical disabilities and the negotiation of leisure constraints. *Leisure Studies, 14*, 17–31.

Hinch, T. D., & Jackson, E. L. (2000). Leisure constraints research: Its value as a framework for understanding tourism seasonability. *Current Issues in Tourism, 3*(2), 87–106.

Hung, K., & Petrick, J. (2010). Developing a measurement scale for constraints to cruising. *Annals of Tourism Research, 37*(1), 206–228.

Hutchinson, S. L., & Kleiber, D. A. (2005). Leisure, constraints, and negative life events: Paradox and possibilities. In E. Jackson (Ed.), *Constraints to leisure* (pp. 137–149). Pennsylvania: Venture Publishing.

Jackson, E. (1997). In the eye of the beholder: A comment on Samdahl & Jekubovich (1997), 'A critique of leisure constraints: Comparative analyses and understandings'. *Journal of Leisure Research, 29*(4), 458–468.

Jackson, E., Crawford, D., & Godbey, G. (1993). Negotiation of leisure constraints. *Leisure Sciences, 15*(1), 1–11.

Jackson, E. L. (2000). Will research on leisure constraints still be relevant in the twenty-first century? *Journal of Leisure Research, 32*(1), 62–68.

Jackson, E. L., & Rucks, V. C. (1995). Negotiation of leisure constraints by junior-high and high-school students: An exploratory study. *Journal of Leisure Research, 27*, 85–105.

Jennings, G. R. (2005). Interviewing: A focus on qualitative techniques. In B. W. Ritchie, P. Burns, & C. Palmer (Eds.), *Tourism research methods - integrating theory with practice* (pp. 99–117). Wallingford: Cabi Publishing.

Kastenholz, E., Eusébio, C., & Figueiredo, E. (2015). Contributions of tourism to social inclusion of persons with disability. *Disability & Society, 30*(8), 1259–1281.

Kitchin, R. (2000). The researched opinions on research: Disabled people and disability research. *Disability & Society, 15*(1), 25–47.

Loucks-Atkinson, A., & Mannell, R. C. (2007). Role of self-efficacy in the constraints negotiation process: The case of individuals with fibromyalgia syndrome. *Leisure Sciences, 29*, 19–36.

McCabe, S. (2009). Who needs a holiday? Evaluating social tourism. *Annals of Tourism Research, 36* (4), 667–688.

Mesquita, S., & Carneiro, M. (2016). Accessibility of European museums to visitors with visual impairments. *Disability and Society, 31*(3), 373–388.

Miller, G., & Kirk, E. (2002). The disability discrimination act: Time for the stick? *Journal of Sustainable Tourism, 10*(1), 82–88.

Nimrod, G. (2008). Retirement and tourism themes in retirees' narratives. *Annals of Tourism Research, 35*(4), 859–878.

Nyaupane, G. P., & Andereck, K. L. (2008). Understanding travel constraints: Application and extension of a leisure constraints model. *Journal of Travel Research, 46*(4), 433–439.

Nyaupane, G. P., Morais, D. B., & Graefe, A. R. (2004). Nature tourism constraints: A cross-activity comparison. *Annals of Tourism Research, 31*(3), 540–555.

Packer, T. L., Mckercher, B., & Yau, M. K. (2007). Understanding the complex interplay between tourism, disability and environmental contexts. *Disability and Rehabilitation, 29*(4), 281–292.

Pearn, M. (2011). Heritage sites: Attitudinal and experimental differences of disabled and able-bodied visitors. In S. Darcy & D. Buhalis (Eds.), *Accessible tourism: Concepts and issues* (pp. 201–213). Bristol: Channel View Publications.

Pennington-Gray, L., & Kerstetter, D. (2002). Testing a constraints model within the context of nature-based tourism. *Journal of Travel Research, 40*, 416–423.

Pritchard, A., Morgan, N., & Ateljevic, I. (2011). Hopeful tourism: A new transformative perspective. *Annals of Tourism Research, 38*(3), 941–963.

Quivy, R., & Campenhoudt, L. (1992). *Manual de investigação em ciências sociais.* Lisboa: Gradiva.

Raymore, L. (2002). Facilitators to leisure. *Journal of Travel Research, 34*, 37–51.

Ren, C., Pritchard, A., & Morgan, N. (2010). Constructing tourism research: A critical enquiry. *Annals of Tourism Research, 37*(4), 885–904.

Richards, V., Pritchard, A., & Morgan, N. (2010). (Re)envisioning tourism and visual impairment. *Annals of Tourism Research, 37*, 1097–1116.

Sedgley, D., Pritchard, A., & Morgan, N. (2011). Tourism and ageing: A transformative research agenda. *Annals of Tourism Research, 38*(2), 422–436.

Silverman, D. (2000). *Doing qualitative data: A practical handbook*. London: Sage Publications.

Small, J., Darcy, S., & Packer, T. (2012). The embodied tourist experiences of people with vision impairment: Management implications beyond the visual gaze. *Tourism Management, 33*, 941–950.

Smith, R. W. (1987). Leisure of disabled tourists: Barriers to participation. *Annals of Tourism Research, 14*(3), 376–389.

White, D. D. (2008). A structural model of leisure constraints negotiation in outdoor recreation. *Leisure Sciences, 30*(4), 342–359.

Williams, P., & Fidgeon, R. (2000). Addressing participation constraint: A case study of potential skiers. *Tourism Management, 21*(4), 379–393.

Yau, M., McKercher, B., & Packer, T. L. (2004). Traveling with a disability: More than an access issue. *Annals of Tourism Research, 31*, 946–960.

Social tourism & older people: the IMSERSO initiative

Diane Sedgley, Claire Haven-Tang and Pilar Espeso-Molinero ⓘ

ABSTRACT

Extant demand-perspective social tourism studies have generally focused on children, families and the disabled whilst older people remain an under-researched group in relation to the personal and social benefits of social tourism initiatives. This study discusses the lack of research on social tourism and older people and takes a demand-side perspective to examine one of the world's most developed and large-scale social tourism schemes for older people, the Spanish Government's IMSERSO programme. A qualitative approach explores the nature of older people's engagement with the scheme and its impact on their wellbeing, through individual in-depth interviews with twenty-seven IMSERSO participants in Benidorm, Spain – a popular IMSERSO destination. The thematic analysis is structured around five themes: social connectivity, the impact of caring responsibilities, a new beginning, escape and practical support. The findings for these IMSERSO participants suggest that, as factors such as caring obligations and hardship experienced in early life are reduced in later life, they have the capacity for increased holiday-taking and social activity and consequently for participation in the IMSERSO scheme. The scheme is found to have a positive impact on their wellbeing by providing opportunities for meeting and interacting with new people, acting as a distraction from ill-health and thus reducing stress, depression and anxiety. However, the article raises questions over the ability of those without strong social networks, including no partner, and those with ongoing caring responsibilities (those who, it could be argued, are most in need of a holiday) to participate in the scheme.

RESUMEN

La existente perspectiva de demanda en los estudios de turismo social generalmente se ha enfocado en los niños, familias y personas con discapacidad mientras que las personas mayores permanecen como un grupo poco investigado en relación con los beneficios personales y sociales de las iniciativas de turismo social. Este estudio discute la falta de investigación en turismo social y personas mayores y toma una perspectiva desde el lado de la demanda para examinar uno de los esquemas de turismo para personas mayores más grandes y desarrollado a nivel mundial, el programa gubernamental español del IMSERSO. Una aproximación cualitativa explora la naturaleza del compromiso de los mayores con el esquema y su impacto en el bienestar, a través de entrevistas en profundidad individuales con veintisiete participantes del IMSERSO en Benidorm, España –un destino popular del IMSERSO-. El análisis temático se estructura alrededor de cinco temas: conectividad social, el impacto de las responsabilidades de cuidado, un nuevo comienzo, escape y apoyo práctico. Los resultados para esos participantes en el IMSERSO sugieren que, como factores tales como las obligaciones de cuidado y privación experimentados anteriormente se reducen al final de la vida, tienen la capacidad para disfrutar de más vacaciones y de actividad social y, consecuentemente, para la participación en el esquema del IMSERSO. Se aprecia que el esquema tiene un impacto positivo en su bienestar al proporcionar oportunidades para el encuentro e interactuación con gente nueva, actuar como distracción de la enfermedad y, así, reducir el estrés, la depresión y la ansiedad. Sin embargo, el artículo plantea cuestiones sobre la capacidad de aquellos que carecen de fuertes redes sociales, incluyendo los viudos, y aquellos que aún tienen responsabilidades de cuidado (aquellos que, podría argumentarse, necesitan en mayor medida unas vacaciones) para participar en este esquema.

RESUME
Les études actuelles sur le tourisme social axées sur la demande se concentrent généralement sur les enfants, les familles, ainsi que sur les personnes handicapées. Les personnes âgées restent cependant un groupe qui n'a pas attiré les chercheurs par rapport aux avantages personnels et sociaux des initiatives de tourisme social. Cette étude analyse l'absence de recherche sur le tourisme social et les personnes âgées et s'intéresse à l'un des programmes de tourisme social les plus développés et les plus importants au monde pour les personnes âgées, en l'occurrence le programme IMSERSO du gouvernement espagnol.

A travers des entretiens individuels approfondis avec vingt-sept participants d'IMSERSO à Benidorm, en Espagne - une destination populaire d'IMSERSO, cette étude explore, par le biais d'une approche qualitative, la nature de l'engagement des personnes âgées dans le projet et l'impact de ce dernier sur leur bien-être. L'analyse thématique tient compte des cinq thèmes: la connectivité sociale, l'impact des responsabilités de soins, un nouveau départ, l'évasion et le soutien pratique. Les participants à IMSERSO suggèrent qu'ils ont la capacité d'augmenter les vacances et l'activité sociale. Par conséquent, ils sont capables de participer au programme IMSERSO lorsque des facteurs tels que les obligations familiales et les difficultés rencontrées au début de la vie sont réduits. Le système a un impact positif sur leur bien-être en offrant des occasions de rencontrer et d'interagir avec de nouvelles personnes, en agissant comme une distraction de la mauvaise santé et en réduisant ainsi le stress, la dépression et l'anxiété. Cependant, l'article soulève des questions relatives à la capacité de ceux qui n'ont pas de réseaux sociaux forts, y compris ceux sans partenaire, et ceux qui ont des responsabilités de prendre soins des autres de façon continue (ceux qui ont le plus besoin de vacances) pour participer au programme.

摘要
现存的从需求角度来看社交旅游的研究一般集中在儿童、家庭和
残疾人，而老年人这块缺乏研究关注，尤其是在社交旅游活动的
个人和社会利益方面。本研究讨论缺乏对社交旅游和老年人的研
究，并以需求方面的观点来审视西班牙政府的计划 IMSERSO，这
是世界上最发达和大规模的老年人社会旅游计划之一。研究采用
定性方法，通过与这个受欢迎的 IMSERSO 目的地西班牙贝尼多
姆的二十七名 IMSERSO 参与者进行深入访谈，探讨老年人参与
计划及其对健康福利的影响。研究通过主题分析法发现五个主
题：社会连通性，关爱责任的影响，新的开始，逃避和实际支
持。这些 IMSERSO 参与者的调查结果显示，由于早年生活中经
历的关怀义务和困难等因素在晚年生活中会减少，因此他们有增
加假期和社交活动，并因此参与 IMSERSO 计划。该计划发现，
通过提供与新人会面和互动的机会，分散对疾病的注意力，从
而减轻压力、抑郁和焦虑，对他们的健康福利产生积极的影响。
然而，这篇文章引发了不少问题，包括那些没有强大社交网络的
老人、　没有伴侣以及那些有照顾责任的人（可能是最需要度假
的人）如何去参加该计划。

1. Introduction

Despite much research on social tourism (see for example McCabe, 2009; McCabe & Johnson, 2013; Minnaert, Maitland, & Miller, 2009) and the accepted benefits of holiday-taking, many people are still unable to participate, hence are excluded from 'social normalcy' (Quinn & Stacey, 2010, p. 31). In Spain, for example, tourism consumption is highly polarised with a relatively small group of people (20%) travelling a lot, whilst 40% of the population have only average participation and 40% of the population are unable to take holidays (Gonzalez & Turegano, 2014). In these circumstances, many Governments, including Spain's, have attempted to widen participation by supporting social tourism initiatives.

An increasing amount of work has been undertaken on the impact of social tourism initiatives on the wellbeing of participants, which has emphasised the long-lasting benefits on happiness, optimism, family life and relationships. Yet much of the focus has been on children, families and the disabled. Fewer studies have focussed on the impact social tourism initiatives can have on the wellbeing of older people; an under-researched group, particularly given the continuous growth of older people as a percentage of the population. Figures from the European Commission (2011) indicate that in the European Union (EU), 18% of Europeans are aged 65 or over, representing 92 million people and by 2060, this will increase to 152.6 million. Spain has experienced significant changes to its population; as the number of people aged over-65 has doubled in less than 30 years and older people account for 16.7% of the country's population. According to Spain's national statistics agency, Instituto Nacional de Estadística, by 2060, 29.9% of the total population will be over-65 (Causapie, 2010). The United Nations (2002) predicts that by 2050, Spain will be the world's 'oldest' country with 40% of its population aged over-60. Consequently, the wellbeing of its older population is going to become increasingly important.

This article is an exploratory study, examining the difference involvement in the Spanish Government's social tourism scheme, 'Instituto de Mayores y Servicios

(IMSERSO) makes to the wellbeing of a group of IMSERSO participants. In doing so, it explores aspects of their home lives to ascertain the extent to which factors important for wellbeing are already in place as well as the difference participation in the scheme makes to their wellbeing. The work is based on a Santander Universities-funded research project, undertaken with support from the University of Alicante, University of the Balearics and Pompeu Fabra University in Barcelona. The paper initially examines extant literature on social tourism, before progressing to tourism literature that focuses on older people and wellbeing. After setting the context for the findings and outlining the methodology, the article then analyses the themes which emerged from the 27 semi-structured biographical interviews with IMSERSO participants.

2. Literature review

Minnaert, Diekmann, and McCabe (2012, p. 29), identify two broad models of social tourism 'host-related' schemes that aim to help host communities and those that focus on the visitors (usually those excluded for health or economic reasons) termed 'visitor-related'. However, they maintain the distinguishing feature of social tourism is that it 'benefits disadvantaged social groups – persons who cannot travel because of financial and/or health reasons or other constraints' and that 'the social aims should be the primary concern for a "social tourism" denomination to stay valid'. They observe that such initiatives, although consistent in terms of the target groups (seniors, disadvantaged young people and families, people with disabilities), adopt a broad range of approaches in terms of the funding arrangements, providers (accommodation, tour operators and services) and the types of holidays offered. Social tourism has a range of participant benefits, particularly for families (Hazel, 2005; McCabe, 2009; Quinn & Stacey, 2010; Sedgley, Pritchard, & Morgan, 2012), with Minnaert et al. (2009) arguing that social tourism provides deprived families an opportunity to increase their family and social capital, widen their social networks, foster positive behaviour, self-esteem and generally increase feelings of wellbeing. Quinn and Stacey (2010) similarly identified increased self-esteem and self-confidence amongst deprived young people; whilst the Family Holiday Association (FHA), has also found that holidays can reduce stress amongst families and give them the confidence to cope with life (FHA, 2015). Sanz, Ferrandis, and Garces (2013) did undertake a study on the relationship between participation in tourism and health, autonomy and social integration amongst older people, it did not focus specifically on social tourism. However, the research did identify a positive relationship between holidaying and self-perceived health and independence and found that those older people who took holidays needed to use social and health care systems less frequently. One of the few studies which has been undertaken on social tourism and older people was the work undertaken by Morgan, Pritchard, and Sedgley (2015) on the impact of a UK based social tourism scheme which found that it offered positive benefits in terms of providing companionship, opportunities for reminiscence, as well as acting as a stimulus for reengagement with physical activity and social interaction. The article concluded by calling for more research exploring the physiological, psychological, social and spiritual impacts of social tourism on the wellbeing of older people.

Despite a proliferation of research on social tourism for families and children, research on the impact of social tourism on the wellbeing of older people has been limited. One explanation for limited research might be that whilst numbers of older people in the EU has increased, the travel industry has focused on the market potential of wealthier and healthier older people, despite the fact that many do not or are unable to travel for a range of reasons, e.g. lack of transportation, poor health, family and caring responsibilities and lack of travel companions (Anderson & Langmeyer, 1982; Blazey, 1986; Guinn, 1980; McGuire, 1984; Romsa & Blenman, 1989; Shoemaker, 2000; Zimmer, Brayley, & Searle, 1995). Tung and Brent Ritchie (2011, p. 332) maintain that the perception of older people as a potentially lucrative market means that 'current research has remained steadfast on investigating their travel motivations, preferences, characteristics, and expenditures'. Concentration on the purchasing power of older people has propagated a perception that they do not need economic support to take holidays. However, the reality is very different – only 41% of seniors within the EU travel (compared to 55% of 25 to 44 year olds) and seven out of ten seniors travel only within their own country (European Commission, 2015). According to data from the Eurostat Toolkit (2017) 'A look at the lives of the elderly in the EU today', the share of older people who travel in Spain is only 39% compared with an EU average of 48.8%. The low levels of economic support and the low participation levels are disconcerting as taking holidays in later life has been shown to have positive physical, psychological, social and spiritual impacts; providing opportunities for activity, social interaction, self-reflection and self-enhancement, thereby enhancing seniors' wellbeing (Dolnicar, Yanamandram, & Cliff, 2012; Hagger & Murray, 2013; Hunter-Jones & Blackburn, 2007; Sanz et al., 2013).

We know little about the difference social tourism initiatives can have on the wellbeing of older people. Despite research indicating that older people are living longer with better physical health than ever before, they are not necessarily getting happier. Problems of isolation, loneliness and depression mean that many older people are living with low levels of life satisfaction (Allan, 2008). As a result, concern about their wellbeing has grown. Whilst there is no accepted definition of what wellbeing is, it tends to be associated with having a pleasurable life, a sense of purpose, independence and dignity (Age UK, 2017). Indeed, in recent years there has been a recognition that older peoples' health is more than the absence of disease but also includes psychological and social wellbeing. Thus whereas traditional models of health emphasised medical, biological and physiological interventions to address the 'dysfunctionality' of ageing (Estes & Binney, 1989; Gullette, 2004; Kaufman, Shim, & Russ, 2004; Phillipson, 2013), more recently, attempts to understand older people's health have been undertaken from a more holistic perspective recognising that psychological wellbeing is also vital and closely related to health (Stephens, Breheny, & Mansvelt, 2015; Wild, Wiles, & Allen, 2013). As Steptoe, Deaton, and Stone (2015, p. 646) state:

> the wellbeing of the elderly is important in its own right and there is suggestive evidence that positive hedonic states, life evaluation and endemic wellbeing are relevant to health and quality of life as people age. Health care systems should be concerned not only with illness and disability but with supporting methods of improving positive psychological states

This small-scale exploratory study therefore aims to identify the impact of the IMSERSO holiday scheme on wellbeing.

2.1. The IMSERSO scheme

In northern Europe, social tourism tends to be the responsibility of charitable organisations whereas in Mediterranean countries, such as France and Spain, the public sector is far more involved (Hunter-Jones, 2011). The Spanish Government's social tourism scheme, IMSERSO is an example of such a publicly funded scheme. IMSERSO began in 1985 and initially offered 16,000 Christmas hotel stays in 19 hotels; it now offers subsidised holidays to older people during the whole of the low-holiday season, October-May. At its peak, in 2009–10, IMSERSO was offering 1,200,000 holidays in 307 hotels (All Parliamentary Group on Social Tourism, 2011). Despite 30% cuts to IMSERSO for the 2012–2013 season (from €102 million to €75 million) (Euro Weekly News, 2012), the scheme continues to receive funding and in 2016/2017 announced a 5% increase in the number of places available (Abaco Advisers, 2015). In the 2015/2016 season, 1.132 million people participated in the scheme (IMSERSO, 2017). Since its establishment, IMSERSO has evolved with increased flexibility, such as offering participants the option of shorter stays and more destinations, including weekend breaks to provincial capitals. Consequently, older people can choose from three forms of holiday: coastal holidays of 15/10/8-day duration, 6-day cultural tours or 5-day nature tours (Abaco Advisers, 2015). To participate in IMSERSO, participants must be resident in Spain, travel during October and June and fulfil criteria, such as being at least 65-years old; retired and in receipt of a public pension or a widower of a pensioner aged 55 or over. For eligible participants, the holiday includes return travel, full-board, double-room accommodation at one of the IMSERSO-selected hotels, travel insurance, entertainment programme (IMSERSO, 2016) and the subsidy amounts to approximately 21% of the total cost of the trip (Cisneros-Martinez, McCabe, & Fernandez-Morales, 2017).

One of the aims of IMSERSO is to create economic benefits, through employment and revenue, for destinations in the low-holiday season. The scheme is regarded as critical in helping to address the problems of seasonality in many Spanish holiday resorts, generating 11,678 direct employment in hotels and 72,574 indirect jobs in Spain (IMSERSO, 2017). As well as the economic rationale for the IMSERSO holiday scheme, it also has the aim of tackling loneliness and improving older people's wellbeing (mental, physical, emotional, psychological, educational or other) by:

- boosting elderly people's social integration
- promoting active ageing
- improving elderly people's quality of life (IMSERSO, 2016)

It is significant that the scheme is administered by the Spanish Institute of Social Services and the Elderly. However, despite the broad aims of the scheme encompassing wellbeing, there has been a preoccupation with examining its economic benefits rather than whether the scheme demonstrably improves older people's wellbeing and thus whether the scheme could be refined to enhance this impact.

3. Methodology

As part of gaining a more rounded understanding older people's wellbeing by, as stated in the Literature Review, moving away from reductionist, medicalised perspectives, many

Governments and gerontologists have begun to focus on the social context of their lives by gathering more personalised subjective reflections (Bond & Corner, 2004; Morgan et al., 2015; Sedgley, Pritchard, & Morgan, 2011). Thus, critical gerontology has emerged to give older people a voice in exploring the complexity of their lives (Jamieson & Victor, 1997). Critical gerontology highlights the need to understand people's personal histories and historical, political, cultural and economic contexts using, for example, biographical research (Bernard & Meade, 1993; Bernard & Scharf, 2007; Holstein & Minkler, 2003; Wilson, 2000). Advocates of critical gerontology within tourism research, Sedgley et al. (2011, p. 430) argue that biographical context:

> allows us to uncover a terrain of life events that help explain the impact of socio-cultural characteristics (e.g. gender, class, ethnicity) and personal factors (e.g. familial, work and friendship networks, where people live, sense of neighbourhood, health, social engagement, consumption patterns) on people's participation and non-participation in tourism ... biographical research also enables us to recognize the wider social and historical changes older people have experienced (e.g. economic and political change) and how these impact on their current circumstances and attitudes.

Thus, within this research, semi-structured biographical interviews explore the participants' life histories and current lives as a way of contextualising the impact of IMSERSO. Interview questions were intentionally broad, covering aspects of life such as marriage, children, and grandchildren as well as the presence of other factors important for determining wellbeing, such as their health, home lives including levels of social connectivity and the nature and frequency of their holiday activity pre-and post-retirement. At the same time, the researchers were mindful of giving the participants a voice in determining the themes and boundaries of the research.

The research team were keen to ensure that the research was carried out in an ethically-sensitive way hence, before commencing the research, the proposed project was scrutinised by the main authors' University ethics committee. This protocol would have been undertaken regardless of the demographic grouping being studied, as Gilhooly (2002, p. 211) observes:

> ... there is nothing special about researching later life, or research with older people. Most older people live independent lives, are self-determining, and are competent to decide whether or not to take part in research. Thus, they should be treated in the same way that one would treat any other adult asked to take part in research ...

However, the ethnographic and biographical elements of the work increased the possibility of eliciting upsetting memories. Consequently, the researchers were conscious of interviewing with empathy and sensitivity.

The primary data collection focused on the Mediterranean coastal resort of Benidorm situated on the Costa Blanca in the province of Valencia. Originally a small fishing village, Benidorm emerged in the 1960s as a large-scale holiday resort. Currently, it is one of the most important tourist destinations in Spain, with approximately 68,000 beds and hotels registering 10 million overnight stays annually (Ivars i Baidal, Rodríguez Sánchez, & Vera Rebollo, 2013) and the second most popular destination (after Andalucia) for IMSERSO scheme participants with 67 hotels participating (IMSERSO, 2017). In total, 27 semi-structured interviews were undertaken in Spanish, alongside the Levante beach in Benidorm where, using purposive sampling, participants were approached by the researcher as

Table 1. Participant profiles.

Pseudonym*	Gender	Age	Marital status	Occupation	Origin
Alberto Andaluz	Male	72	Married	Civil Servant	Madrid
Elena Alcántara	Female	69	Married	Housewife	Madrid
María López	Female	68	Married	Housewife	Village Guadalajara
Valle Solas	Female	67	Married	Housewife	Basque Country
Camilo Suarez	Female	69	Married	National guard	Basque Country
Carmen Martínez	Female	74	Widow	Housewife	Murcia
Mª Luisa López	Female	72	Married	Office Clerk	Murcia
Carmen (Margarita) Gómez	Female	86	Widow	Housewife	
Laura Cisneros	Female	86	Widow	Housewife	
Marisa Molina	Female	70	Married	Housewife	Madrid
Javier Chacón	Male	71	Married	Fitter	Madrid
Antonia Martinez	Female	71	Widow	Entrepreneur and dressmaker	Renteria (Basque Country)
Rosa Pascual	Female	62	Married	Shoe repairwoman	north of Spain
Luis Sánchez	Male	67	Married	Woodworker	north of Spain
Carmina Fernandez	Female	70	Widow	Housewife	Cantabria
Aurora (Isabel) Figueroa	Female	68	Married	Office Clerk	Madrid
Marcelino López	Male	70	Married	Bus Driver	Madrid
Carmela Blasco	Female	72	Married	Housewife	Granada
Antonio Romero	Male	77	Married	Civil Servant	Granada
Pilar Gómez	Female	70	Married	Teacher	Motril (Granada)
Pepe García	Male	77	Married	Farmer	Motril (Granada)
Encarnación Sánchez	Female	75	Married	Labourer	La Zubia (Granada)
Clara Barreros	Female	77	Married	Housewife	La Zubia (Granada)
Ana Nieto	Female	75	Married	Dressmaker	Madrid
Eusebio Martín	Male	78	Married	Decorator	Madrid
Aurora Sánchez	Female	72	Widow	Housewife	Madrid
Alicia García	Female	73	Widow	Housewife	Madrid

*Women in Spain do not change their family name after marriage. Married women have different family names than their husbands.

they walked along the promenade in the area. The participants ranged in age from 62 to 86 and of these, 20 were female and only 7 were male. This female predominance is understandable given that that the majority of participants in the scheme are female (IMSERSO, 2017). All were married or had been married (7 were widowed, the rest still married). Interestingly, 20 of the respondents had very little experience of travel before retirement. The characteristics of the respondents are summarised in the table above (Table 1).

The interviews were all recorded and transcribed into English so that thematic analysis could be undertaken. Five themes were identified: social connectivity, the impact of caring responsibilities, a new beginning, escape and practical support. Quotes from the narratives have been used to present the results of the analysis and to reflect the personalised insights of the participants. It is worth noting that, as the interviews are full of regional and colloquial words, translation was difficult. In those instances, the decision was taken to interpret rather than directly translate the respondent's words to ensure the essence of their meaning was not lost. Some respondents chose pseudonyms to protect their identity.

4. Results and discussion

4.1. Social connectivity

In order to contextualise the significance of participating in IMSERSO holidays on well-being, the interviews began by exploring the participants' home life in terms of family ties and social support networks. The strength of these ties was striking, reflected by high levels of regular face-to-face or telephone family contact and that six of the respondents still had adult children living at home. An insight into the high levels of family contact came from Encarnation:

> I have four children, the youngest lives at home, he is 45-years old; he doesn't even have a girlfriend. All of them live in my village ... Some of my children come every day and others come less but, if they cannot come, they phone me.

When Valle was asked how often she sees her children she said, laughing, 'too often'. Similarly, Elena explained that the level of contact with her family was such that 'I sometimes look for a little time of loneliness for myself'. Likewise, Maria Luisa, when asked if she ever felt lonely, stated 'Lonely? Quite the opposite'.

The closeness of these respondents towards their children both reflects the fact that in Spain, young people stay at home until a relatively late age, a trend exacerbated more recently with the economic crisis and high levels of unemployment (Carrasco & Rodriguez, 2000) and that their cohort grew-up under Francisco Franco, who enforced a set of social structures designed to preserve the traditional role of the family. Under his regime, motherhood was identified as a woman's main role and a national and societal obligation (Valiente, 2013). The idea of *perfecta casada* (the perfect housewife) and *angel del hogar* (angel of the home), emphasised the value of personal sacrifice, particularly in relation to raising children (Reyes, 2015). Such values have had lifelong implications for participants, as evidenced later in this paper.

As well as the importance of close family relationships, many of the participants also referred to the importance of close friends:

> We have very good friends; there are a couple of such good friends that are like siblings to us. We used to live in the same building and the doors of our houses were always open ... the boys were coming back-and-forth in both houses. They moved away from Madrid, but our friendship is still very strong, after 40 years (Javier).

Many of these friendships and social ties stem not only from being close neighbours, as in Javier's case, but also from the social activities of the participants, such as membership of the Association of Salesian (Friars) Alumni, Cultural Associations, local associations for retired people and Yoga groups. The significance of these social networks in terms of life satisfaction is clear, as illustrated by Alicia:

> I do everything, I don't want to stop. I do Tai Chi, a memory class, crafts. Every day I go to the [community] centre and any other day I go out with my friends, we are always in touch.

The Church, as might be expected in a Roman Catholic country like Spain, was at the centre of many respondents' social lives, providing both social connectivity and spiritual wellbeing, as reflected by Maria:

> We have a prayer group that is wonderful, I feel very welcome, it's superb people, we pray, we sing, we praise, we have service on Sundays and catechism on Thursdays.

Many respondents' comments confirm that the factors determining older people's wellbeing, such as relationships with family and friends, social contact, religion/spirituality, community engagement and leisure, already feature strongly in these participants' lives (Bond & Corner, 2004). Indeed, López García, Banegas, Graciani Pérez-Regadera, Herruzo Cabrera, and Rodríguez-Artalejo (2005) found that only a small proportion of

Spain's elderly lack frequent social relationships. This common factor in many partici-pants' lives is reassuring as high levels of social contact have been found to reduce the risk of depression, even in those older people with physical health problems (López García et al., 2005). Conversely, other research has found that deficiencies in social relationships are even associated with an increased risk of developing heart disease and strokes (Valtorta, Kanaan, Gilbody, Ronzi, & Hanratty, 2016). The major challenges which older people face in later life, such as bereavement, retirement, illness of a close partner and taking on the role of carer, can be alleviated with the support of family and friends. However, a lack of social support is linked with an increase in mortality and poor health. The dominant theme of strong social and family ties in this work differs from that of Morgan et al. (2015), who found loneliness to be a recurring theme amongst English participants of a UK social tourism scheme. In the UK, loneliness and isolation has long been identified as a major problem, where 51% of all people over-75 live alone and 5 million identify television as their main form of company (Age UK, 2014).

However, even in Spain, despite the strong social connectivity and the undeniably enjoyable formal and informal activities on offer, feelings of loneliness and loss are not eradicated, as Carmina describes:

> I walk every day for an hour-and-a-half listening to music. After that, I go with my friends in the afternoon and we play Parchisi and chat. We go to the Catholic Centre. Once a month we have a dance, and besides that we do parties and daily trips ... [but] I do feel lonely. When you are out, you feel accompanied but then, when you close the door of your house, you need somebody to talk to about things. I was 44 when I was widowed, and you never get used to it. When my children leave the house I'm lonely.

Significantly, the majority of interviewees in this research were married, which perhaps reflects the ease of participating in social activities, including holidays, if one has a partner (Age UK, 2017). In Age UK's latest study on wellbeing, they found that older people in the highest wellbeing group were more likely to be married whilst a third of those on the lowest wellbeing group were widowed. A recent study by Sommerlad, Ruegger, Singh-Manoux, Lewis, and Li (2017) even went so far as to suggest that married people are less likely to develop dementia as marriage encourages both partners to exercise more, eat more healthily and smoke and drink less, as well as offering more opportunities for social engagement. Whilst the IMSERSO scheme does acknowledge the challenges facing those who are widowed by allowing those aged 55, rather than 65, to participate in the scheme and also allows participants to take a companion with them who doesn't necessarily have to be a pensioner, the extent to which people in this category take up this opportunity is not clear. IMSERSO's annual report (IMSERSO, 2017) on the scheme only provides data on the gender, age, income and levels of satisfaction of the participants, suggesting that a broader analysis would be valuable to understand their participants' homes lives, including levels of social connectivity and marital status, in order to identify the any common charac-teristics of those groups who do not participate in the scheme.

4.2. The impact of caring responsibilities

Whilst celebrating the time they now had to spend with friends and participate in social activities, many respondents recalled how their earlier lives had been characterised by caring for children followed by caring for elderly parents:

I have been all my life at my home caring for all of mine ... I have looked after everyone, my mother with oxygen, my father with dementia. My father-in-law spent 28 years at home with me and some weeks also his sister was at home, all those seniors I had in my house and a sister, she was in a wheelchair for a year and three months. All have passed through my hands, with no-one to help me (Maria).

Neither my husband nor I had any siblings ... We had elderly parents, caring for them first and then, my children were born, then we got my in-laws ... Eventually after all those years, I am travelling with my husband (Carmen).

Intense caring duties meant that pre-retirement holidays had been difficult. Twenty participants stated that they travelled very little pre-retirement or, if they did travel, they had had to take elderly relatives with them. Only recently had they been able to holiday as couples, free of caring responsibilities. María Luisa describes how caring for both her mother and mother-in-law impacted on her holidays and her perception of holidays:

My mom was in my house, she passed away five years ago, and I also had my mother-in-law for 45 years in my house, she died when she was 100 years old; I took care of both, imagine! ... [On holiday] I had to rent a house and I went with the two grandmothers, my mother and my mother-in-law, just imagine that scene. They were not holidays at all.

The huge caring responsibilities, once again reflects the strong sense of family loyalty and personal sacrifice found amongst many older women in Spain, where there is almost a moral duty to keep elders within the family home (Alberdi, 1999). Carrasco and Rodriguez (2000, p. 51) observe that the significance of women's unpaid caring has been so extensive that it has served 'as a shock absorber for adversity and economic crisis'.

At the same time, research has also shown that familial obligation towards ageing parents has shielded many Spanish elders from the detrimental impact of loneliness and, subsequently, poor physical and mental health. Accordingly, there are concerns that, as Spanish society changes, with increased female employment and children moving away, the supportive relationship between children and the elderly will be eroded. As Zunzunegui, Beland, and Otero (2001, p. 1091) state, the 'mismatches in the expectations of elderly parents and their children may arise with increased frequency and the deleterious effects of poor support on the health of the older people may be more readily detectable'. Given that Spain is experiencing a period of economic and social transition, this societal shift raises questions as to whether the older people in this research will receive the same levels of care they dedicated to others.

Currently, the strong sense of family obligation and solidarity continues in this cohort of women so, whilst many are no longer caring for older relatives, they now have extensive responsibilities for their grandchildren:

My husband goes to pick up my grandchildren, he brings them from school and we eat at home, I have six or seven sitting on my table everyday (Maria).

We have a daughter. We see her every day because she comes home for lunch. She has a child and we see him every day as well. I take him to school and then I pick him up and he comes home for lunch with us (Antonio).

Previous research in Spain found that when parents struggle with work and family commitments, the extended family (mainly grandmothers) often intervene to look after children, especially early infant child-care (Carrasco & Rodriguez, 2000; IPSOS Mori, 2011).

Extending the scheme to include inter-generational family breaks could address caring commitments for those with grandchildren, however, further research would be required to examine the implications for the overall aim of IMSERSO, not least the ability of older people to participate in low-season holidays and the release of older people from caring duties.

4.3. A new beginning

Generally, many respondents had been liberated from the major care responsibilities of elderly relatives, children and grandchildren and thus able to enjoy their new-found leisure time with a clear conscience, knowing that they did their utmost to care for relatives. Valle, for example, describes how she treated her aunt 'like a marchioness' and that 'She came like a pauper and went like a queen'. This sense of having fulfilled familial obligations is also evident in the words of Elena:

> We have had [dependents], a very difficult situation, we had to travel every two months to Córdoba [300 km] for two years to assist my mother-in-law, and then my parents ... we couldn't go anywhere ... we have served them very well and now we can enjoy the satisfaction of having helped them and be at peace with ourselves.

Interestingly, Warren and Clarke (2009) also found a sense of liberation amongst older people, particularly the sense of freedom resulting from declining family and work commitments. Within this research, Encarnation described her enthusiasm for travel having been freed from caring roles 'because before we were never able to see beyond the door of the house.' Whilst twenty of the participants stated that they had travelled very little pre-retirement, since retirement, nineteen described travelling extensively, taking 2–3 trips a year. For some older people, these holidays 'represent not simply the time of their lives, but also time for their lives' (Dann, 2001, p. 10). Therefore, whilst IMSERSO does allow older people who have caring responsibilities to take that person with them or, to take their own carers, it would be valuable to know how many older people in these categories are participating in the scheme.

4.4. Escape

Many respondents described the sense of freedom resulting from their participation in IMSERSO, as well as the sense of escape it provides from worries at home, where they often feel anxious and tend to dwell on their physical and mental health issues:

> Here you relax. I don't know why, but you relax ... I suffer from anxiety and here I feel great, I wake up and I have nothing to do, even my knee is better here (Ana).

> Doctors? I didn't even think about it. I bring my book and that's it. No anxieties at all. We come here to relax (Carmen).

> You really disconnect, I come back very relaxed (Marisa).

In Clara's case, being on holiday distracts her from worries connected to her husband's health and mortality:

> When I come away with IMSERSO, I'm very relaxed. At home, my nerves don't allow me to sleep, but here, after a couple of days here, I sleep very well. I don't have nerves at all ... Often

at home I have anxiety problems, but when I get onto the bus I start to feel better. Here I sleep at siesta time and I also sleep at night … Since my husband got sick I am scared that something is going to happen to him. Sometimes I even cry, but it doesn't happen when I am here. I've been all my life with him, and I don't know what I'll do if something happens to him; but here it does not happen. As soon as I start to unpack at home, I start to feel nervous again.

Similarly, Aurora describes how taking part in the IMSERSO trips has allowed her to overcome bouts of depression, particularly after her husband's death:

I started to travel with IMSERSO with my husband. At first, I didn't want to but then I started to warm to the idea. When my husband passed, my sons and the doctor encouraged me to come back; I was falling into a depression. It suited me very well, I got distracted, I met some great people … The first time, I came alone with a married couple of friends, but lately I come with lady-friends.

The positive atmosphere clearly distracts the participants from their own or partner's ill-health and mortality:

I think it helps to make you feel better … When you are doing well we forget a lot of things. It feels good … The atmosphere that you find here, to meet people, be with other people, get to know how older people are, talking with people from other places …. . it makes you learn, and you make very good friends (Valle).

From Valle's perspective, it is the social aspect of the holidays, particularly meeting new people from other regions of Spain, which allows her to forget her troubles:

I like IMSERSO, you get to know many people, we have made many friends …. . I mean, you make very good friendships, I'm very friendly, I get along with people very easily and here there is a very good atmosphere with music every night. I like it.

Such sentiments support previous work (Corlyon & La Placa, 2006; Hazel, 2005; McCabe, 2009) which found that holidays can reduce stress, allow time for relaxation and recovery and the rebuilding of emotional strength. McCabe (2009) highlighted that holidays have the potential to reduce stress and allow people to cope with ill-health and chronic conditions. Similarly, Morgan et al. (2015, p. 11) found that one of the most beneficial aspects of subsidised holidays for older people was that it punctuated individuals' everyday routines, extended their social worlds and represented an opportunity to escape everyday responsibilities, thus enabling 'clients to contextualise their problems, re-evaluate their lives, confront negative self-images and develop coping strategies to increase their emotional resilience to recent life transitions such as reduced socio-economic circumstances, bereavement and increased ill health and physiological change.'

The perceived advantage of Benidorm for many respondents is that it is a vibrant, all-year round destination. Antonia describes less satisfactory experiences when visiting other Spanish destinations during the low season:

Once I went to Ibiza in March … but everything is closed; where can you go? … There are no buses running … every move has to be done by taxi, and taxis are very expensive. Later on, I went to La Manga and exactly the same.

The attraction of Benidorm in the low season is that there are plenty of people, hotels and shops are open and the transport system is operational. Elena, for example, gains inspiration from the liveliness of the resort and the many elderly people around her who are managing their health and not giving up:

> It rejuvenates you, you see the way people move … look at that lady, how old she is and how she moves … It therefore stimulates you, you say, 'well, we cannot be sitting at home.

This attitude supports the work of Jerrome (1989) who has observed how the attitudes of older people's peer group to ageing provide 'normative guidelines' for older people facing the challenges of later life. The findings also support Minnaert et al. (2009) who found that social tourism can provide respondents with an opportunity to reflect on their lives and identify areas where change is needed; similarly, McCabe (2009, p. 682) found that family holiday schemes can give participants a 'fresh sense of perspective on problems'; whilst Morgan et al. (2015) found that holidays provided older people on UK social tourism schemes an opportunity to re-evaluate their lives and confront negative self-images. Generally, many of these studies have emphasised how the benefits of social tourism extend beyond the immediate holiday experience into the everyday lives of participants. There is clearly an opportunity for IMSERSO, to explore the impact of the scheme on richer features of participants' wellbeing, as well as collect data for their annual reports on participants' satisfaction with their holiday.

4.5. Practical support

As well as the purported psychological benefits, such schemes are important to older people for the practical help, guidance and support offered:

> They [IMSERSO] really help. Sometimes you don't have a group in your location, but they combine groups with another city and you just have to join the group in that place. The girl or the boy that goes with the group helps you to join, they take you to the boarding area, tell you the gate, they help you with everything (Luis).

> You don't have to worry about a thing, it's all organized. When we used to travel on our own, we had to seek accommodation, search for parking for the car, find tickets to go this place or the other. Not anymore. We get everything done for us (Pilar).

> IMSERSO helps you a lot. There are seniors who want to travel but they don't know which plane they have to take; where they have to sit; and where to go to pick up their boarding pass and they [IMSERSO employees] explain everything. IMSERSO personnel wait for you, tell you which plane; they bring you to the hotel, they help you with everything (Rosa).

There is no doubt that both the psychological and practical benefits of the IMSERSO scheme explain why, since retirement and with freedom from work and caring responsibilities, many of the respondents not only continue to take part in IMSERSO (20 had been on IMSERSO trips before) but also take advantage of other social tourism schemes in Spain organised by regional governments, city halls, retiree associations, church groups, or professional guilds. Elena, for example, takes full advantage of all travel opportunities:

> With IMSERSO we do everything we can … then we also travel with Alcorcon City Council, which also organize trips, the Community of Madrid, we take whatever suits us at that time. We think travelling is culture, each site is a different thing, it opens up life. I love it.

The fact that both national and local Governments in Spain continue to support social tourism schemes suggests a broad base of support, which is unusual when many

Governments are moving towards anti-interventionist and liberal capitalist systems, less concerned with equality and social justice (Higgins-Desbiolles, 2012; McCabe, Minnaert, & Diekmann, 2012). Higgins-Desbiolles (2012, p. 65) argues that the hegemony of neoliberalism means that many Governments have been forced to celebrate the benefits of tourism purely around a discourse of economics.

5. Conclusions

Our results highlight how, for these participants when caring obligations, mothering duties and hardship experienced in early life is reduced it leaves space for increased holiday-taking and social activity; which subsequently reduced loneliness. This work examined further how the social tourism delivered by IMSERSO was identified with overcoming anxiety, countering depression, providing a distraction from ill-health and reducing stress. The work has reinforced the findings of other research on social tourism which has shown it can provide opportunities for meeting and interacting with new people, improving participants' mental and physical health, providing practical support and enhancing their subjective wellbeing.

It is clearly difficult to differentiate the effects produced because of Spain's social context from those resulting purely from social tourism. However, respondents identified clear benefits and related these to IMSERSO and the examples cited by the respondents were intimately associated with their experiences of social tourism. Being away from home, detached from caring obligations and with opportunities to create new friendships, further reduced stress, loneliness and had a positive and additional impact on wellbeing.

The research has also shown that unhappiness in later life is not inevitable and that initiatives can be introduced which foster wellbeing and social inclusion. Unfortunately, much of the support for older people within Europe is modelled on the traditional medical model of service delivery, based on medical notions of decline and 'care'. Hence the focus on interventions in the lives of older people tend to ignore the key part that more holistic policies based on seeing, valuing, and responding to the whole person can play in maintaining confidence, capacity, and societal contribution.

This research does highlight the positive impact of the scheme on wellbeing amongst those participating. However, it does acknowledge that many of these participants already had many of the important enabling factors needed for wellbeing in place such as social capital, a partner, freedom from caring responsibilities and reasonable levels of health. Whilst IMSERSO has put in place provisions to ensure that those who may be more socially isolated, due to widowhood, caring responsibilities and those with carers can participate, a more detailed analysis of the groups are taking up this opportunity would provide insight into its value. Also, in the spirit of critical gerontology, a more holistic understanding (people's personal histories and historical, social, cultural and economic contexts, including levels of social connectivity), of the lives of those who are participating might provide improved insight into causes of low participation for groups. This could inform a move towards greater involvement from health, social care and community care organisations in the scheme and improve the involvement of vulnerable groups in the scheme.

To conclude, this research has highlighted how tourism, if viewed from a non-commercial perspective, can be used as part of a holistic approach to older people's wellbeing.

Social tourism should thus be included in public health and wellbeing policy discourses on empowerment and inclusion as it can contribute to improving people's quality of life (La Placa & Corlyon, 2014). It is thus concerning that tourism research has given scant attention to older people's wellbeing and the extent to which holidays might enhance self-esteem, education, cross-cultural interaction, spiritual growth and solidarity (Higgins-Desbiolles, 2012). However, at a time when European Governments seem to believe that individuals, rather than the state, should take responsibility for their own lives (Monbiot, 2016), the Spanish Government should be admired for continuing to support the scheme.

Acknowledgements

The authors wish to acknowledge the support of the University of Alicante, University of the Balearics and Pompeu Fabra University in Barcelona.

Disclosure statement

No potential conflict of interest was reported by the authors.

Funding

The work was funded by an internal staff mobility fund granted by Santander Bank, as part of the Cardiff Metropolitan University and Santander Universities agreement.

ORCID

Pilar Espeso-Molinero ⓘ http://orcid.org/0000-0001-5876-3906

References

Abaco Advisers. (2015). *IMSERSO goes from strength to strength*. Retrieved from Abaco Advisers website: http://www.abacoadvisers.com/spain_explained/blog/life-spain/imserso-goes-strength-strength
Age UK. (2014). *TNS loneliness omnibus survey*. London: Author.
Age UK. (2017). *A summary of Age UK's index of wellbeing in later life*. Retrieved from Age UK's website: https://www.ageuk.org.uk/globalassets/age-uk/documents/reports-and-publications/reports-and-briefings/health--wellbeing/ageuk-wellbeing-index-summary-web.pdf
Alberdi, I. (1999). *La nueva familia Española* [The New Spanish Family]. Madrid: Taurus.
Allan, J. (2008). *Older people and wellbeing*. Retrieved from the Institute of Public Policy Research website: http://www.ippr.org/images/media/files/publication/2011/05/older_people_and_wellbeing_1651.pdf
All Parliamentary Group on Social Tourism. (2011). *Giving Britain a break: Inquiry into the social and economic benefits of social tourism*. London: TSO.
Anderson, B., & Langmeyer, L. (1982). The under-50 and over-50 travellers: A profile of similarities and differences. *Journal of Travel Research, 20*, 20–24.
Bernard, M., & Meade, K. (Eds.). (1993). *Women come of Age – perspectives on the lives of older women*. London: Edward Arnold.
Bernard, M., & Scharf, T. (Eds.). (2007). *Critical perspectives on ageing societies*. Bristol: The Policy Press.

Blazey, M. A. (1986, October). Research breathes new life into senior travel program. *Parks and Recreation*, pp. 55–56.

Bond, J., & Corner, L. (2004). *Quality of life and older people*. Maidenhead: Open University Press.

Carrasco, C., & Rodriguez, A. (2000). Women, families and work in Spain: Structural changes and new demands. *Feminist Economics*, 6, 45–57.

Causapie, P. (2010). *Active ageing. Lychnos: Notebooks of the Fundacion General CSIC* [Adobe Digital Editions version]. Retrieved from Fundacion General CSIC website http://www.fgcsic. es/lychnos/en_EN/forum/active_ageing

Cisneros-Martinez, J. D., McCabe, S., & Fernandez-Morales, A. (2017). The contribution of social tourism to sustainable tourism: A case study of seasonally-adjusted programmes in Spain. *Journal of Sustainable Tourism*, 25, 1–9.

Corlyon, J., & La Placa, V. (2006). *Holidays for families in need: UK policies and guidelines – final report of the family holiday association*. London: : Policy Research Bureau.

Dann, G. M. S. (2001). Targeting seniors through the language of tourism. *Journal of Hospitality and Leisure Marketing*, 8(1), 5–35.

Dolnicar, S., Yanamandram, V., & Cliff, K. (2012). The contribution of vacations to quality of life. *Annals of Tourism Research*, 39(1), 59–83.

Estes, C. L., & Binney, E. A. (1989). The biomedicalization of aging: Dangers and dilemmas. *The Gerontologist*, 29(5), 587–596.

European Commission. (2011). *The 2012 ageing report: Underlying assumptions and projection methodologies*. Retrieved from European Commission, Economic and Financial Affairs Directorate website: http://ec.europa.eu/economy_finance/publications/european_economy/ 2011/pdf/ee-2011-4_en.pdf

European Commission. (2015). *Calypso: Tourism for all*. Retrieved from European Commission website: http://ec.europa.eu/growth/sectors/tourism/offer/calypso/index_en.htm

Eurostat. (2017). *A look at the lives of the elderly in the EU today*. Retrieved from European Commission website: http://ec.europa.eu/eurostat/cache/infographs/elderly/index.html

Euro Weekly News. (2012). *Almeria ready to welcome elderly*. Retrieved from Euro Weekly website: https://www.euroweeklynews.com/news/costa-blanca-north/itemlist/tag/IMERSO

FHA. (2015). *Impact report* 2015. Retrieved from Family Holiday Association website: http://www. familyholidayassociation.org.uk/research/impact-report/

Gilhooly, M. (2002). Ethical issues in researching later life. In A. Jamieson & C. R. Victor (Eds.), *Researching ageing and later life* (pp. 211–225). Buckingham: Open University Press.

Gonzalez, R. P., & Turegano, S. M. A. (2014). Consumo turístico y desigualdad social en españa. *PASOS Revista de Turismo y Patrimonio Cultural*, 12(1), 29–51.

Guinn, R. (1980). Elderly recreational vehicle tourists: Motivations for leisure. *Journal of Travel Research*, 19, 9–12.

Gullette, M. M. (2004). *Aged by culture*. Chicago: University of Chicago Press.

Hagger, C., & Murray, D. (2013). Anticipating a flourishing future with tourism experiences. In S. Filep & P. Pearce (Eds.), *Tourist 647 experiences and fulfilment: Insights from positive psychology* (pp. 186–202). Abingdon: Routledge.

Hazel, N. (2005). Holidays for children and families in need: An exploration of the research and policy context for social tourism in the UK. *Children and Society*, 19(3), 225–236.

Higgins-Desbiolles, F. (2012). Resisting the hegemony of the market: Reclaiming the social capacities of tourism. In S. McCabe, L. Minnaert, & A. Diekmann (Eds.), *Social tourism in Europe: Research and practice* (pp. 53–66). Bristol: Channel View.

Holstein, M. B., & Minkler, M. (2003). Self, society, and the "new gerontology". *The Gerontologist*, 43(6), 787–796.

Hunter-Jones, P. (2011). The role of charities in social tourism. *Current Issues in Tourism*, 14(5), 445–458.

Hunter-Jones, P., & Blackburn, A. (2007). Understanding the relationship between holiday taking and self-assessed health: An exploratory study of senior tourism. *International Journal of Consumer Studies*, 31, 509–516.

IMSERSO. (2016). *Imserso tourism program*. Retrieved from the Institute for Older Persons and Social Services Website: http://translate.google.co.uk/translate?hl=en&sl=es&u=http://www.imserso.es/imserso_01/envejecimiento_activo/vacaciones/&prev=search

IMSERSO. (2017). *Annual report, 2016. Ministry of health, social services and equality* [Informe Anual IMSERSO 2016. Ministerio de Sanidad, Servicios Sociales e Igualdad]. Retrieved from the Institute for Older Persons and Social Services Website: http://www.imserso.es/interpresent3/groups/imserso/documents/binario/informeanual2016.pdf

IPSOS Mori. (2011). *Children's well-being in UK, Sweden and Spain: The role of inequality and materialism*. Retrieved from UNICEF United Kingdom website: http://www.unicef.org.uk/Documents/Publications/IPSOS_UNICEF_ChildWellBeingreport.pdf

Ivars i Baidal, J. A., Rodríguez Sánchez, I., & Vera Rebollo, J. F. (2013). The evolution of mass tourism destinations: New approaches beyond deterministic models in Benidorm (Spain). *Tourism Management, 34,* 184–195.

Jamieson, A., & Victor, C. (1997). Theories and concepts in social gerontology. In A. Jamieson, S. Harper, & C. Victor (Eds.), *Critical approaches to ageing in later life* (pp. 175–187). Buckingham: Open University Press.

Jerrome, D. (1989). Virtue and vicissitude: The role of old people's clubs. In M. Jeffreys (Ed.), *Growing old in the twentieth century* (pp. 151–165). London: Routledge.

Kaufman, S. R., Shim, J. K., & Russ, A. J. (2004). Revisiting the biomedicalization of aging: Clinical trends and ethical challenges. *The Gerontologist, 44*(6), 731–738.

La Placa, V., & Corlyon, J. (2014). Social tourism and organised capitalism: Research, policy and practice. *Journal of Policy Research in Tourism, Leisure and Events, 6*(1), 66–79.

López García, E., Banegas, J. R., Graciani Pérez-Regadera, A., Herruzo Cabrera, R., & Rodríguez-Artalejo, F. (2005). Social network and health-related quality of life in older adults: A population-based study in Spain. *Quality of Life Research, 14*(2), 511–520.

McCabe, S. (2009). Who needs a holiday? Evaluating social tourism. *Annals of Tourism Research, 36* (4), 667–688.

McCabe, S., & Johnson, S. (2013). The happiness factor in tourism: Subjective well-being and social tourism. *Annals of Tourism Research, 41*(1), 42–65.

McCabe, S., Minnaert, L., & Diekmann, A. (Eds.). (2012). *Social tourism in tourism: Theory and practice*. Clevedon: Channel View.

McGuire, F. A. (1984). A factor analytic study of leisure constraints in advanced adulthood. *Leisure Sciences, 6,* 313–326.

Minnaert, L., Diekmann, A., & McCabe, S. (2012). Defining social tourism and its historical context. In S. McCabe, L. Minnaert, & A. Diekmann (Eds.), *Social tourism in Europe: Research and practice* (pp. 18–31). Clevedon: Channel View.

Minnaert, L., Maitland, R., & Miller, G. (2009). Tourism and social policy: The value of tourism. *Annals of Tourism Research, 36*(2), 316–334.

Monbiot, G. (2016). *How did we get into this mess? Politics, equality, nature*. London: Verso.

Morgan, N., Pritchard, A., & Sedgley, D. (2015). Social tourism and well-being in later life. *Annals of Tourism Research, 52,* 1–15.

Phillipson, C. (2013). *Ageing*. Cambridge: Wiley.

Quinn, B., & Stacey, J. (2010). The benefits of holidaying for children experiencing social exclusion: Recent Irish evidence. *Leisure Studies, 29*(1), 29–52.

Reyes, J. R. (2015). *Spain - the changing attitudes in Spain - gender, cohabitation, family, definition, development, and family*. Retrieved from JRank Website: http://family.jrank.org/pages/1622/Spain-Changing-Attitudes-in-Spain.html#ixzz3UdUsPh00

Romsa, G., & Blenman, M. (1989). Vacation patterns of the elderly German. *Annals of Tourism Research, 16,* 178–188.

Sanz, M. F., Ferrandis, E. D., & Garces, J. (2013). Functional health benefits for elderly people related to social tourism policy promotion. *International Journal of Multidisciplinary Social Sciences, 1,* 1–8.

Sedgley, D., Pritchard, A., & Morgan, N. (2011). Tourism and ageing. A transformative research agenda. *Annals of Tourism Research, 38,* 422–436.

Sedgley, D., Pritchard, A., & Morgan, N. (2012). 'Tourism poverty' in affluent societies: Voices from inner-city London. *Tourism Management, 33*(4), 951–960.

Shoemaker, S. (2000). Segmenting the mature market: 10 years later. *Journal of Travel Research, 39*, 11–26.

Sommerlad, A., Ruegger, J., Singh-Manoux, A., Lewis, G., & Li, G. (2017). Marriage and risk of dementia: Systematic review and meta-analysis of observational studies. *Journal of Neurology, Neurosurgery and Psychiatry*, Published Online First: 28 November 2017. doi:10.1136/jnnp-2017-316274

Stephens, C., Breheny, M., & Mansvelt, J. (2015). Healthy ageing from the perspective of older people: A capability approach to resilience. *Psychology & Health, 30*(6), 715–731.

Steptoe, A., Deaton, A., & Stone, A. A. (2015). Subjective wellbeing, health and ageing. *The Lancet, 385*(9968), 640–648.

Tung, V., & Brent Ritchie, J. R. (2011). Investigating the memorable experiences of the senior travel market: An examination of the reminiscence bump. *Journal of Travel and Tourism Marketing, 28*, 331–343.

United Nations. (2002). *World population ageing 1950–2050*. Retrieved from the United Nations Website: http://www.un.org/esa/population/publications/worldageing19502050/pdf/preface_web.pdf

Valiente, C. (2013). Gender equality policymaking in Spain (2008–11): Losing momentum. In B. N. Field & A. Botti (Eds.), *Politics and society in contemporary Spain: From Zapatero to Rajoy* (pp. 179–196). New York, NY: Palgrave Macmillan.

Valtorta, N. K., Kanaan, M., Gilbody, S., Ronzi, S., & Hanratty, B. (2016). Loneliness and social isolation as risk factors for coronary heart disease and stroke: Systematic review and meta-analysis of longitudinal observational studies. *Heart*. Retrieved from http://heart.bmj.com/content/early/2016/03/15/heartjnl-2015-308790

Warren, L., & Clarke, A. (2009). 'Woo-hoo, what a ride!' Older people, life stories and active ageing. In R. Edmondson & H. J. von Kondratatowitz (Eds.), *Valuing older people: A humanist approach to ageing.* (pp. 233–248). Bristol: Policy Press.

Wild, K., Wiles, J. L., & Allen, R. E. (2013). Resilience: Thoughts on the value of the concept for critical gerontology. *Ageing & Society, 33*(1), 137–158.

Wilson, G. (2000). *Understanding old age: Critical and global perspectives*. London: Sage.

Zimmer, Z., Brayley, R. E., & Searle, M. S. (1995). Whether to go and where to go: Identification of important influences on seniors' decisions to travel. *Journal of Travel Research, 33*, 3–10.

Zunzunegui, M. V., Beland, F., & Otero, A. (2001). Support from children, living arrangements, self-rated health and depressive symptoms of older people in Spain. *International Journal of Epidemiology, 30*(5), 1090–1099.

Index

For Product Safety Concerns and Information please contact our EU
representative GPSR@taylorandfrancis.com
Taylor & Francis Verlag GmbH, Kaufingerstraße 24, 80331 München, Germany

www.ingramcontent.com/pod-product-compliance
Lightning Source LLC
Chambersburg PA
CBHW081543220326
41598CB00036B/6539